The Buffalo Tail: A Memoir, 1921-2010

By Daniel T. "Bud" Kelly, Jr.

To Walter - a friend from Las Vegas and an asset to El Castillo. Bud
Daniel T Kelly Jr
8/26/2016

Edited by Don J. Usner

© 2012 Daniel T. Kelly, Jr.
All Rights Reserved.

No part of this publication may be reproduced, stored in a retrieval system, or transmitted, in any form or by any means, electronic, mechanical, photocopying, recording, or otherwise, without the written permission of the author.

First published by Dog Ear Publishing
4010 W. 86th Street, Ste H
Indianapolis, IN 46268
www.dogearpublishing.net

ISBN: 978-1-4575-1104-2

This book is printed on acid-free paper.

Printed in the United States of America

Dedication

For their love, tolerance, and companionship
during my lifetime journey I dedicate this book
to my parents,
Daniel Thomas Kelly and Margaret Gross Kelly;
my wife,
Jeanne Wise Kelly;
and to my children,
Susan Ellis, Thomas Linton, Robert Walker,
and Pamela Wise

Acknowledgments

My gratitude to Kay Carlson for her patience in deciphering my handwriting and typing up my journal entries. I'm grateful also to Marigay Graña for reviewing the manuscript and offering suggestions for improvement, and to my daughter Pamela and Betsy Joyce for designing the book cover. Special appreciation and thanks to Don Usner for his editorial skill and diligence in shepherding the manuscript through the publication process. These friends made this book possible.

Table of Contents

A NOTE ON THE TITLE xi

1921-1934—BOYHOOD IN SANTA FE 1

1934-1943—PREP SCHOOL AND HARVARD 12

PORTSMOUTH PRIORY 12

HARVARD UNDERGRADUATE DAYS 20

1943-1945—WORLD WAR II 34

1946—POSTWAR DUTY: ABERDEEN AND AUSTRIA 50

1947-1950—HARVARD REVISTED AND A RETURN
  TO NEW MEXICO 62

HARVARD BUSINESS SCHOOL 62

RETURNING TO NEW MEXICO AND WORKING WITH
  GROSS KELLY & CO. 66

1951-1952—KOREAN DUTY 71

1953-1959—MARRIAGE, LAUNCHING A CAREER
  AND DOMESTICATION 84

1953: COURTSHIP AND MARRIAGE 84

1954: A SEARCH FOR A CAREER 88

1955: A FULL LIFE 90

1956: ARRIVAL OF THE TWINS 94

1957: FOUNDATIONS FOR GROWTH 96

1958: FAMILY LIFE 98

1959: URANIUM AND NEW INTERESTS 100

1960-1969—THE SIXTIES: LEARNING, ADJUSTING
   AND ACHIEVING 104

1960: BUSINESS AND FAMILY 104

1961: COMMUNITY INVOLVEMENT 105

1962: PLAYING COWBOY 109

1963: NEW MEXICO AMIGOS 113

1964: FAMILY TRAUMA 116

1965: ACQUIRING THE AGENCY 118

1966: THE NEW REGIME 121

1967: OPERA HOUSE FIRE 124

1968: DRAMATIC TIMES 127

1969: THE END OF THE SIXTIES 130

1970-1982—PROSPERITY AND CHALLENGE 132

1970: THE CALDERA 132

1971: A YEAR OF TRANSITION 134

1972: A BUSINESS GOAL ACHIEVED 137

1973: TRAUMA AND RELEASE 139

1974: RELATIVE TRANQUILITY 143

1975: THE RIFT 145

1976: REORGANIZATION 149

1977: INTERMEZZO 151

1978: THE WAYFARERS 154

1979: A POTPOURRI—HOME AND NEPAL 157

1980: COLD WARS—CHANGES IN THE SKI INSURANCE INDUSTRY 160

1981: Vague Discontentment	163
1982: Passing the Baton	165
1983-1993—TRANSITIONS AND FAREWELLS	172
1983: The New and the Old	172
1984: Interludes in the Routine	173
1985: Mother's Passing	176
1986: The Phoenix	179
1987: Changes in the Air	181
1988: The Conflict	183
1989: Cow Creek and the Erection Party	184
1990: The Revolving Stage	187
1991: Birthdays and Milestones	193
1992: The Battle	196
1993: Adjusting	199
1994-2000—NEW PASTURES	203
1994: The Age of Adult Puberty	203
1995: Diary Reflections	206
1996: Excursions and Interludes	210
1997: Resolutions	214
1998: Travel and Trauma	218
1999: Le Fin de Siècle—Relative Calm	221
2000-2011—SEIZING THE DAYS	232
2000: Cuba and France Revisited	232
2001: Penance and Absolution	238
2002: Flight for Life, and Historical Travel	243
2003: Frustrations and Indecision	248
2004: Letters from Caroline	250

2005: A Living Treasure 254

2006: A Fiftieth Milestone 257

2007: Venture Capitalism and a Pilgrimage 260

2008: Financial Frustration 265

2009: A Reminiscent Phone Call 268

2010: Looking Back 271

2011: Postscript 274

A Note on the Title

I'm calling this memoir "The Buffalo Tail" to connect it with *The Buffalo Head*, my father's autobiography and history of the family business, Gross Kelly and Company, a commission and trading company active in the Southwest from 1867 until 1954.

Our family's life revolved around Gross Kelly and Co. from as early as I can remember. As an adult, I was intimately involved in the company's activities and thus had a close association with my father, beyond our family relationship. I learned much from him and I admired and emulated him in many ways, although we were very different people. I took over that business and served briefly as its president before closing it in 1954.

The demise of Gross Kelly & Co. represented the end of an era, as did the passing of my father in 1973. I was fortunate to catch the tail end of that era, during which our family not only witnessed momentous historical changes, but also participated in making those changes happen. I followed that trail through much of the twentieth century and then entered new realms in business life, navigating my way into a new century in Santa Fe. This is my story.

1921-1934

BOYHOOD IN SANTA FE

My mother's maid, Ethel Kerr, announced my arrival into the world on March 22, 1921, in a white brick bungalow on the corner of Federal Place and Lincoln Avenue in Santa Fe, New Mexico. My father had come home for lunch when Ethel declared, "Mr. Kelly, you have a son."

Ethel, one of the Gross family's black servants, had accompanied my mother, Margaret Gross, a new bride on her move west from St. Louis. Mother's father, Jacob Gross, and my dad's father, Henry W. (Harry) Kelly, were business partners operating a mercantile warehousing operation, Gross Kelly & Co., with branches in Colorado and New Mexico. The company was an outgrowth of an older business that accompanied the railroad as it progressed west in the late 1800s.

I never knew Grandfather Gross. He had retired from the New Mexico scene but kept abreast of the company's progress from St. Louis. Grandfather Kelly was the resident New Mexico partner and actively ran the business from Las Vegas, New Mexico, where my father, Daniel T. Kelly, Sr., was raised.

My recollections of my Grandfather Kelly are vivid. I liked him a lot and wanted to impress him. I held him in awe. I remember him as a practical joker. At Christmas we visited my grandparents in Las Vegas. We would put up stockings at the two main fireplaces in the house. On Christmas morning, we'd find the stockings located near the Christmas tree were stuffed with conventional stocking gifts—candy, fruit, small things. The other stockings were hung over the fireplace in the formal den in a less-used part of the house. These we found stuffed with chunks of coal. At the bottom of the each stocking, however, was a gold coin of considerable value.

Grandfather hosted Christmas dinner in the large, formal dining room. The maids, Petra and Marcella, wore their white uniforms. Grandfather always carved the turkey. He relished giving me the less attractive parts of the bird, the gizzard and the tail, which he called "the Pope's nose."

I hunted blackbirds with Grandfather Kelly. I recall he was an excellent shot and enjoyed bringing back a full game bag, but I remember him best in his business persona. At 7:30 a.m. every work day, his faithful assistant and driver, McDaniel, would arrive in a big, shiny car and take him to work at the Gross Kelly office/warehouse complex in the Las Vegas rail yard. The building stood adjacent to the sumptuous Castaneda hotel, one of the railroad hotels in the Fred Harvey chain. At the end of his work day, around 3:30 p.m., Grandfather would be driven back home in the shiny car. When he arrived at the house, he went immediately to his liquor cabinet, where he would practically inhale two jiggers of bourbon. He then took a nap. This workday habit dated back to his younger days, when whisky was part of the frontier business routine. As he grew older, the habit stuck, but the effects of the liquor had a more pronounced effect on him.

I remember one time coming across Grandfather taking a different kind of siesta. He was crashed out in the fabled hide and wool warehouse at the far end of the company complex of buildings. This wooden, odiferous building housed sacks of wool, bails of dry cow hides and sheep pelts, and the tanning vats where green cow hides were treated. In decades past, the room had held stacks of buffalo hides. The manager of the warehouse was Matias, the "hide-house foreman." My grandfather was partial to this part of the business, and his lord-and-servant relationship with faithful Matias was longstanding and close.

It was a hot day. I was exploring around the company facilities and decided to call on Matias, whom I liked. I proceeded to the hide house and entered the building. There, lying on a full eight-foot-long sack of wool, I found Grandfather Kelly, fast asleep. He was dressed in his usual, formal business attire—suit, vest and necktie, with a gold watch chain crossing the vest to his watch pocket. The gold chain pulsated up and down as he breathed. I had caught him in his hideaway, sleeping off the effects of too much whisky. Matias cautioned me to keep my discovery a secret. (Shortly after the incident, Grandfather Kelly became ill and died at a sanitarium in California.)

Notwithstanding his weakness for whisky, Grandfather Kelly was a remarkable man who reigned over an awesome business empire. He was born in Leavenworth, Kansas in 1858 and apprenticed in 1873 with Otero, Sellar and Company, a commission and freight forwarding business advancing westward with the railroads. In 1881, at the age of 26, he became a partner in the successor company, Gross Blackwell Co., which in 1879 moved to Las Vegas, New Mexico, along with the railroad. By 1902 the company was known as Gross Kelly & Co. and enjoyed a reputation for being the largest business of its kind in the Southwest. There were branches of the company in Albuquerque, Pecos, Tucumcari, and Epris, New Mexico, and in Trinidad, Colorado.

In addition to being a multi-product warehousing operation, Gross Kelly was involved in the wool and livestock business and timber operations in New Mexico and the Republic of Mexico. Grandfather held financial interest in Arnot Wool Company and the Jackson Cattle Company in Las Vegas, the Maxwell Timber Company in Cimarron, New Mexico, and the Kelly, Kemper and Kitch Cattle Co. of Denver, Colorado. He was involved in telephone companies in the cities of Chihuahua and Durango, Mexico. (His life story is detailed in my father's book, *The Buffalo Head*.)

In 1913 Gross Kelly & Co. opened a branch in Santa Fe, and in 1919 my father moved there as an employee of the new office. The St. Louis partner in the business, Jacob Gross, occasionally came out to New Mexico on business visits. It was during one of Grandfather Gross's summer visits to Las Vegas that my parents became acquainted. They married in 1916 in Normandy, Missouri, a suburb of St. Louis. The newlyweds initially lived in Trinidad, Colorado, where Dad was employed in another company branch. After serving in the army in World War I, in 1919 he was transferred to Santa Fe to manage the company's operations in Santa Fe and north into Colorado. My parents settled in the home on Lincoln Avenue.

I was one of six Kelly children, which came in two "litters." Their first child was Henry Warren II (Hank), born in 1917 and named after Dad's father. Then came Caroline Linton in 1919, and me, Daniel T. Kelly, Jr., in 1921. (I'm not sure when they started calling me Bud, but it must have been early on. My father, too, had been called Bud, or Buddy, as a child, but not as an adult.)

Marie Ellis (Yei), born in 1923, was the last in my parent's first litter, which was the formative one for me. The second brood came later, with Mark Gross Kelly (1929) and William Booker Kelly (1934). These last two in many ways represented a different generation, raised by battle tested and more tolerant parents.

My few recollections of my childhood life on Lincoln Avenue are hazy, but I remember we had a black German shepherd named Doodle. I recall one time when two of my siblings and I, the three oldest children in the family, dressed up in Indian costumes: khaki colored outfits with red tassels sewn on them and war bonnets with chicken feather decorations. I recall running a red Crayola along the wall as I maneuvered down the stairs from my room—a creative act that drew no accolades.

We moved to Palace Avenue in 1925, when I was four years old, and our home on that street has remained the focal point for my life ever since. This is still home for me now, in 2010. The trees, garden and house exterior remain much the same, but the world around has changed immensely, from those days of horses and Model-A Fords to these days of instant communication, space exploration and a one-world, interdependent economy.

Santa Fe in the 1920s and early 1930s still seemed like a colonial town, with a distinct separation between the colonizing Anglos and the Spanish-American, or Mexican, residents. For the most part, the recently arrived Easterners were the *patrones*, the Rajas, while the natives served as maids, garden keepers and day laborers (*trabajadores*). Spanish was the language of this working class, although the *gringos*, or Anglos, knew enough to give orders or make their wishes understood. The natives were staunch Catholics and their employers were mostly Protestants who bore a tolerant bias towards their Catholic employees and neighbors.

Bertha Bonderont, another émigré from Mother's St. Louis, was the Negro cook in our home. Cleofus Lobato was the Hispanic laundress. A dignified and handsome woman, Cleofus wore a black shawl over her head and spent her days in the laundry washing and ironing. She would arrive early in the morning on foot, bringing her daughter Josefita and niece Incarnation, along as helpers and house apprentices. Both of these young girls grew up and spent their lives as key domestics in our household. Mother gave the orders and oversaw all household affairs, but she left the details and duties to the maids. Often, the maids' workday would end after a late dinner party, and they would spend the night in the "maids' room" or walk in the dark to their own home—only to recommence the day's routine early next morning.

Mother enjoyed supervising the gardener and chore man. We had access to an acequia for irrigating, which kept the grounds green and watered a vegetable garden. We enjoyed the food grown there and milk from a resident cow. Only when Luterio, the chore boy, was off on a drunk or seeking penance as a Penitente did my father venture near a hay bale or a milk pail. Manual labor was not his bag. His philosophy was to find a more efficient method of maintenance than his own labor, and to invest in it.

I remember the acequia that flowed around the perimeter of our property and the hard labor required to distribute the water where it was needed. To simplify this arduous task, my father devised an elaborate waterworks. He had the main ditch cemented to reduce evaporation and installed head gates and valves to direct the flow of water. He made sure all planting areas were level. All this preparatory work was done just before the summer irrigation period commenced. Putting the whole system together cost him considerable money, and then on the very day the new system was to be inaugurated, city officials declared the ditch system non-operative. The historic era of acequia irrigation, begun with Spanish colonization centuries before, ended on Palace Avenue.

Viewing human relations from today's viewpoint, life in the '20s and '30s was tough for the working class and the rewards modest. However, this comparison is not altogether relevant. In my family's relationship with our staff, there was a strong feeling of loyalty, dependence, and appreciation on both sides. The amenities at 531 Palace Avenue, as well as the excitement of parties and social events, gave spice to our employees' lives and offered them some variation from the ordinary. At the same time, having close associations and daily interactions with the local people was rewarding and interesting for my family. Exposure to the Hispanic and Indian ways of life—regional diet, social customs and religious celebrations—afforded us a broader perspective that warmed and nurtured our lives.

As Josefita came of age, she became our cook. Fat and jolly, she liked preparing red chile with lamb shanks, white and blue corn tortillas with red or green chile, sopaipillas and honey, and a traditional spinach-like vegetable called *verdolagas* (purslane). Not only did Josefita enjoy cooking; the delectability of the results and compliments from the consumers gave her great pleasure. She was also superstitious, believing in certain supernatural phenomena, especially *brujas* (witches) and ghosts. On occasion, when she was afraid, she relished having one of us young ones snuggle up to her ample and warm body for mutual reassurance.

One time when my parents were on an extended Mediterranean cruise, Josefita, who was of Spanish-Indian derivation, and Bertha, a Missouri black, were left in charge of my siblings and me. For some reason, the idea of seeing the silent movie *Dracula* appealed to our two guardians. We proceeded to the

Paris movie theater on San Francisco Street, where we kids sat on either side of Josie and Bertha. Dracula (Bela Lagosi) was so horribly frightening that we all spent part of the movie on the floor of our aisle, cowering from the vampire. For several days afterwards, we children were bedmates, sleeping with either Bertha or Josie; they were as frightened of a possible visit by Dracula as we were.

My brother Hank was my mentor throughout my childhood. I followed him like a puppy, and he included me in his activities. He was imaginative, tolerant, and the obvious leader of his group of friends. As a result of spending so much time with Hank and his pals, I was exposed to the point of view of kids about five years older than me.

Hank was small and wiry. He wore eyeglasses and some said he looked like a sissy. He was fresh meat for the Mexican groups that made up most of the student body at St. Michael's, the school he attended. Fights were constant, and Hank, often the solo *gringo* boy on the playground, was picked on and trounced upon by the gangs, so my father decided Hank should learn to box. Connie Mac, an ex-welterweight professional, coached him. Hank became a skilled boxer and crushed his oppressors, earning their respect.

Boxing tournaments were popular in the schools at the time, probably because of the fame of Jack Dempsey, the "Manassas Mauler," a world boxing champion who hailed from Colorado. Hank became the titleholder in his weight class and kept up the sport in college, ultimately becoming captain of the Harvard College boxing team.

With Hank's initial coaching, I became interested in boxing, but I didn't have a formal coach. Thinking that just copying my older brother would suffice for training, I entered a tournament. In my first fight, my opponent was a 10-year-old, redheaded Mexican who had a reputation as a tough boxer. I entered the smoke-filled ring with misplaced confidence, but somehow managed a first swing to his belly that knocked out his wind. I won the match and concluded there was nothing to this sport. I was supremely confident I would win my next fight, but this time my opponent knew how to box and he pummeled me unmercifully and won the match. At that point I gave up boxing and only used it in self-defense against the gangs that later plagued me at school.

My education was eclectic. In Santa Fe there were primary public schools and parochial schools. The student body at the parochial schools comprised mostly Mexican Catholics, while the public schools were more ethnically mixed and nonsectarian. St. Michael's was a private Catholic school for boys.

There also were other private schools supported by recently arrived, college-educated Eastern Anglos who had "discovered" Santa Fe. These schools were generally faddish, however; they were short lived and attended only by the more affluent gringos.

My parents valued education. My mother was inquisitive and well read. My father was not a scholar, but he was a strong supporter of Harvard College, his alma mater. For some reason, I never attended kindergarten or the first grade. Instead, I was privately tutored by a great lady, Mrs. White, a former public school teacher who lived down the street on Palace Avenue. I think I was switched from left-handed to right-handed writing under her guidance. She was strict and very proper, but warm and likable.

I spent the second grade in Catron School, where my teacher was Miss Earl, Mrs. White's sister. Most of her class of about thirty students was Anglo. Miss Earl was a bespectacled, no-nonsense, professional schoolmarm who earned the students' respect.

For the third grade, I transferred to St. Michael's, a Catholic day and boarding school for boys, established in Santa Fe in 1859 by the De La Salle Christian Brothers, a teaching order of French origin. The Brothers, who spoke French-accented English and little or no Spanish, experienced considerable frustration and trauma as they tried to teach in English to the predominantly Spanish-speaking students. We students, around nine years old, deliberately exacerbated the situation. Talking, throwing spitballs, scuffling, and generally baiting the poor Brothers led to frequent temper outbursts from these essentially patient and holy men. I was no saint. I recall several times being banished to the crawl space under the schoolroom floor for my misbehavior. That punishment, or having to kneel on the floor with our knees pressing into the wall molding for long periods of time, did little to reform us. The Brothers occasionally used our heads as blackboard erasers.

For the fourth grade, my parents tried a new tack, enrolling me in Miss Wheelwright's private, all-Anglo school on Canyon Road. The classes were of mixed age and gender. There I was exposed to a group of girls who were older and more physically advanced than me.

Puppy-like, the boys followed the lead of this bevy of young sirens. It was they who hatched the idea of stealing candy from a nearby Canyon Road grocery store. At recess time, the kids would enter the store. While boys diverted the owner's attention, the girls stuffed Mr. Goodbars, Milky Ways, and Hershey bars into their bloomers. We carried out this larceny about three times before the owner got wise to us when a candy bar dropped on the floor from a culprit's underwear.

The thieving ring broken, my father insisted I confess and apologize to the owner and repay him for the pilfering. This I did, while my father waited outside. The punishment cured me of stealing.

For fifth, sixth and seventh grades, I was back in public school with a regular group of primary grade girls and boys. Apparently, I was just a regular student. I do remember being very fond of one of the girls. She was pretty and, more importantly, her father was a former rodeo rider and a saddle maker, which piqued my interest because of my love of all things "cowboy."

For the eighth grade my parents sent me back with the Christian Brothers at St. Michael's. At this age the boys were more disciplined, but the academics were as marginal as before, as were my grades. My parents decided on a drastic change for me: Eastern boarding school.

My parents were both Catholics. Religion was a strong factor in their lives. My grandmother's parents had converted to Catholicism. As is typical of converts, they were more devout than people born to that faith. My mother's family, on the other hand, came from a background of intellectual Catholics, and some of her relatives were nuns, priests, and bishops.

Santa Fe was a Catholic city, so adapting to its religious culture came naturally to my parents. Mother joined the Altar Society and Dad often served as a secular advisor to the resident archbishop. We children played our part in the scene, receiving religious instruction and participating in the church's routines, although many of my Anglo friends were not Catholic. Academically, the Catholic schools were adequate, but the predominately Mexican student body set the teaching pace. Because of the anemic academic environment as well as cultural factors—and, probably, for parental self-preservation—my siblings and I all eventually ended up attending Catholic boarding schools.

Not all of my education took place in schools. In fact, I learned some of my life lessons and useful skills during horse riding adventures, "cowboying" with ranchers, attending Indian and Hispanic religious ceremonies, and going on archeological field trips and other adventures in places distant from Santa Fe.

I was introduced to horses at an early age. My mother had a favorite riding mare named Yegua (Spanish for "mare"). We kept Yegua in the stable in the rear of our Palace Avenue property, along with two unexciting but reliable plugs, Maxine and Anheiser, who permitted us children to climb all over them. From our back yard, you could "ride to Denver" without interference from fences or roads.

When I was about twelve years old, I met Bud Davis, who owned the Sanguilla Ranch at Sapello, some twelve miles north of Las Vegas, New Mexico. The Davis family was prominent and politically active with the Republican Party. Stephen B. Davis, Bud's father, served in Herbert Hoover's cabinet. Bud went to Dartmouth College. When I met him, he had just returned from a stint with the United Fruit Company in Central America and was working as the Sanguilla ranch manager.

For some reason, Bud took a fancy to me and became like a big brother, uncle, and great friend. From him I learned cowboy ways of riding, roping and handling cattle. He taught me to appreciate the fun of cowboy sports, to drive a truck and to shoot rabbits and prairie dogs. We "fox hunted" coyotes, following horseback as his two Irish wolf hounds frantically chased for a kill. He would rope yearling calves and have me mount the frantic beasts, only to see me get bucked off once he let go of the rope.

In exchange for all his mentoring, I think I gave Bud an excuse to depart from the somewhat pedantic routine of livestock raising. I spent part of two summers with Bud's family. I adored his charming Canadian wife, Eleanor, whom he had met while she was convalescing from tuberculosis in the Las Vegas, New Mexico area. I went with them to post-prohibition parties given by the neighborhood ranchers. Bud had an enormous effect on me, and my friendship with him and his family is a cherished memory.

I had an altogether different experience of horses and cowboying when I "worked" on the Gross Kelly ranch south of Las Vegas, on the Pecos River near Dilia, New Mexico. The lifestyle was the antithesis of that on the Davis ranch. Ranch life at Dilia was an unglamorous affair. The cowboys, bachelors all, lived in very primitive conditions at the headquarters adobe: no electricity, no indoor plumbing, no refrigerators, no radios, no automobiles. The ranch work was strictly a horse-and-wagon operation. It was ranching the old way.

Tom Acord was the Anglo foreman of the company ranch. His hired hands, the Alarid family cowboys, had grown up ranching. Their father, Ruperto Alarid, in his early seventies, was the camp cook. He had been an open-range cowboy in eastern New Mexico and West Texas. He was a dignified, weather-beaten man who, despite his age, could ride and herd cattle, drive a chuck wagon, and cook simple but palatable meals. I was his charge.

Ruperto taught me to sit tall in the saddle and to always be on guard for a surprise from the mount. He spoke in broken English with a Texas cowboy accent, telling me innumerable stories about incidents he witnessed on the open range. One time, he treated me when a tick became lodged in my ear. Tobacco juice was the "tick antivenin" he administered; he spit the juice into my ear with some hesitation but with great accuracy. I rode with Ruperto on the chuck wagon as he drove the four-horse team from roundup camp to roundup camp. He was an oracle from an age just ending and I learned a lot from him.

I was also exposed to Indian culture while growing up. My family often attended feast day celebrations and dances at the pueblos of northern New Mexico and eastern Arizona. Among our favorites were the Santo Domingo Corn Dance, the Shalako House Blessing ceremony at Zuni, and the Snake Dance and Bean Dance at the Hopi mesa pueblos.

My father was well acquainted with several of the archeologists of the day, such as Jesse Nussbaum and Alfred V. Kidder, and leading historians like France V. Sholes. As a result, I was privileged to know and converse with a number of these scholars. I also visited archeological research sites. I remember touring Mesa Verde with my father and the park director, Jesse Nussbaum, before much of it was open to the public. Exploring the ruins and cave dwellings in Frijoles Canyon in Bandelier National Monument was also an exciting adventure for our family, well before access became highly controlled. Nothing was off limits to us there. We could explore as we wished.

We participated in New Mexico's Spanish heritage through the Catholic feast and saint days celebrated in Santa Fe. Of course the annual Santa Fe Fiesta was an especially glamorous and festive occasion. In those days, wearing costumes for fiesta was almost a requirement. One year for the pet parade I led around the plaza a leashed, pet turkey I had raised. The poor bird was never the same after this experience.

Another activity I recall in my early youth was "Penitente hunting" with my parents and their friends. Penitentes were members of a Hispanic religious brotherhood that practiced self flagellation and other forms of penance requiring the shedding of blood for remission of sins. The season of Lent was the time for their ceremonies. One clear, cold night when we were "hunting" at Abiquiú, we hid behind a hillock near the *morada*, the chapel where the Penitentes were gathered. The doors suddenly opened and candlelight poured out in the darkness. Four men pulling as a team brought forth the Carreta de la

Muerte. Mounted on the cart was a wooden skeleton with a poised bow and arrow. Attendants, swirling hollow bones attached to ropes, created a weird and spooky "bull roar" sound. We watched the procession proceed down the hill. (Legend had it the jarring of the cart on the bumpy path would dislodge an arrow from la Muerte's bow, striking one of those pulling the carreta.)

Although I never witnessed it, the Penitente hermanos sometimes reenacted a crucifixion. To this day, they still hold prayer ceremonies in the moradas, although the self flagellation is prohibited by Church officials.

1934-1943

PREP SCHOOL AND HARVARD

Portsmouth Priory

In 1934 I enrolled in Portsmouth Priory School in Rhode Island—a far cry from New Mexico and its culture of Penitentes, pueblo dancers, and tobacco-spitting cowboys. I started out by repeating the eighth grade and then spent five years as a boarder, graduating in 1939. The transition was extreme: from Mexican, high desert New Mexico to Rhode Island, the smallest of states, and to Aquidneck Island near Newport. No family, no mountains, no Mexicans, no Indians, no horses, no chile—nothing was familiar. The boys, all gringos, were smart, and the dress code demanded coats and neckties. In contrast to New Mexico, the Rhode Island weather was wet, with the sky nearly always overcast. It seemed that water was everywhere.

Portsmouth Priory (now Portsmouth Abbey School) was a Catholic college-preparatory boarding school operating in conjunction with a Benedictine monastery. It stemmed from its mother house, the Fort Augustus Abbey on Loch Ness in Scotland, an English Benedictine congregation. The headmaster, Father John Hugh Diman, had founded St. Georges, an Episcopal boarding school in nearby Newport, before he converted to Catholicism. Ordained a priest and monk at Fort Augustus Abbey, Father Diman was tapped as the first headmaster of the Portsmouth school.

The school property, formerly a country estate, sat atop a sloping hill bounded by the shores of Narragansett Bay. The meadows of the rustically landscaped estate had been converted into athletic fields separated by rows of large leafy trees. Turtles, frogs, pheasants and birds made their home in a creek bed overgrown with trees and brush. Here the Revolutionary War battle of Cory's Lane had been fought.

The manor house, a stately three-story frame structure, complete with three porches, stood at the heart of the school. It overlooked the bay and Prudence Island, upon which stood a lighthouse. The contrast between arid New Mexico and ocean-bounded Rhode Island constantly amazed me. The school property included almost a mile of relatively untouched beach.

The school program was structured around six grades, known as forms, running from seventh through twelfth. Secular lay masters and ordained monks taught English, Latin, French, math, physics, chemistry, history and athletics. Central to the schedule was chapel, with a morning Mass and evening compline (prayers at the end of the day). All services were geared to the monastic devotional ritual of the Benedictine Order.

Housemasters, all of them bachelors, supervised the dorms, although sixth-form seniors (student prefects) maintained order. Discipline was strict. The English headmaster, Dr. Bateman, used caning as a penalty for serious infractions.

During my time at Portsmouth Priory, the student body consisted of only about one hundred boys. Small class size and a low student-teacher ratio allowed considerable personal attention for each pupil. Although life was highly regimented, students were generally happy and kept occupied by studies and by athletics—basketball, baseball, football, ice hockey, tennis, squash and, in season, sailing. Winter weekends we went sledding and in spring we explored nature on the beaches and in the woods surrounding the school. If we were lucky, visiting parents treated us to delicious meals at excellent restaurants in nearby Newport. Girls were an unknown species, but as time went on, curiosity and body chemistry generated the urge to learn more about things feminine.

The faculty was dedicated and well informed. Naturally, we liked some teachers and merely tolerated others, but we made them all the brunt of our roguish humor. Like the students, the teachers led lives that revolved fairly strictly around the school calendar, but unlike us, they could taste the pleasures of "normal" society on weekends, when they could leave the campus.

"Bunny" Cooke, a diminutive Englishman, stood out among the characters on the faculty. He taught math and was the master of St. Benedict's dormitory. He detested the students' habit of using scrap paper to figure out math problems. He had a temper, which he usually—but not always—restrained. When he did rage, he seemed to grow in stature from four-foot-twelve to a frightening height. He liked his sherry and left nighttime dormitory discipline up to the prefects while he "hit the sack."

Bunny's faculty pal Bill Wimsett taught English. Standing seven feet tall, he coached basketball and baseball. He could suspend a basketball in one hand, palming it, as they say. When he was a baseball umpire, the strike zone for him was somewhere between the chest and the head of the batter. Later in his career, he became a tenured English professor at Yale.

Dr. Lally, the history teacher, had graduated from Johns Hopkins and obtained his Ph.D. in England. Lally was a saintly, totally dedicated, if absent-minded, professor whom everyone respected and who engendered a love of history in his students.

My sixth-form year, I served as the prefect of the "New" dorm under housemaster W. Griffith Kelley. Kelley's field was English literature. He was a charming, urbane gentleman who had previously been headmaster of Canterbury School in Connecticut. He often invited me to visit with him, once the younger boys were supposedly safely asleep in their beds. I cherished my exposure to his knowledge of books, authors, current affairs, and life in general. Tragically, after I had graduated he was brutally mauled and robbed by thugs in New York City and never recovered from the beating.

Becoming familiar with the traditions and religious ceremonies of the Benedictine Order during my five years at Portsmouth had a lasting influence on my life. The order was established by St. Benedict of Nursia (480-553 C.E.). Among its monasteries is the Italian Abbey of Monte Cassino. In 596, St. Augustine of Monte Cassino introduced the Benedictine order into England, one of the first countries outside of Italy in which Benedictine life was firmly planted. The order brought education, letters, and learning to the country. Britain's universities and great cathedrals were largely cultural outgrowths of Benedictine monasteries.

The Reformation, initiated in Henry VIII's reign, resulted in almost total elimination of these historic, influential and wealthy monasteries. The Benedictine Order in England went underground and migrated to Europe. When the order was permitted to return to Britain in the nineteenth century, monasteries were reestablished there. Downside Abbey in England and Fort Augustus Abbey in Scotland furnished the initial cadre of monks that established Portsmouth Priory monastery and the school in Rhode Island.

(Incidentally, several of the Portsmouth monks, seeking a more isolated and spiritual setting, later established the Monastery of Christ in the Desert near Abiquiú, New Mexico, north of Santa Fe).

During prep school summer vacations, I usually returned to Santa Fe. It was something of a culture shock to go back and forth between these very different worlds. I recall one year in particular when I was returning from prep school in June. The Corpus Christi procession happened to be going on when I arrived. As the procession flowed from St. Francis Cathedral to and around the plaza, it stopped at selected spots where the participants venerated the sacred host contained in the monstrance, which was transported on a bier by selected men of the parish. At intervals, a squad of men with rifles would fire a salute. It seemed very foreign and somewhat pagan to me, having just returned from Rhode Island. Nevertheless, I joined in the procession.

During summer breaks, David Davenport and I took some adventurous, extended trips. One year we attended the World's Fair in California. En route we stopped in a canyon near Reno, Nevada. We parked the car and put out bedrolls on the arroyo sand and went to sleep. During the night, a cloudburst filled the dry canyon with a roaring flood. We rushed to the car and were able to reach the bank of the arroyo in time. The next morning, the spot where we had parked the car was strewn with large, man-killing boulders.

The comic ending to the trip to California was David's fascination with Sally Rand's Nude Ranch. He could not see enough of those rouge beauties.

David and I also took a very ambitious trip to Mexico City. We were 18 years old. We dressed like cowboys. David's red hair fascinated the Mexican señoritas. I would introduce him as *"el hijo de Tom Mix"* (the "son of Tom Mix," the popular movie cowboy of the day).

On the day we visited the sacred chapel of the Virgin of Guadalupe, David, an Episcopalian, was besieged by women who pinned ribbons and medals of the Virgin on his shirt front. That afternoon we attended an Episcopal tea party held by U.S. expatriates. The guest of honor was an Episcopalian priest visiting from the United States—coincidentally, the same priest that had christened David eighteen years before, on Long Island. He was saddened when he saw David bedecked with the ribbons of the Virgin, as he assumed David had left the Episcopal Church and had converted to Catholicism.

The Mexican trip ended with a chilling experience for me. David's father had become very ill, which required David to fly home. It took all our money to buy his plane ticket, leaving me with and an unpaid hotel bill and a car to drive home to Santa Fe. Fortunately, my father had given me a letter of introduction to the National City Bank in Mexico. (This was the bank operated by William Blaine Richardson Jr., the father of Bill Richardson, who would later become the governor of New Mexico.) With refreshed finances, I paid the hotel bill and departed alone for the drive back to the United States.

I was somewhat foggy from driving when I pulled into a small, country town just south of Ciudad Victoria. It was market day and the road was crowded with tequila-drinking laborers bearing machetes. They surrounded my car and seemed to be threatening me. I panicked and hit the throttle, breaking through the crowd. I saw two or three men bounce off the hood of the car to fall into the roadside ditch.

I sped off, terrified the gang of men might catch up with me and filled with fear I had injured or killed people. My imagination went wild. I was sure I was being pursued by a Mexican police car. I stopped in Ciudad Victoria, where I knew of an American hotel owner. I limped into the lobby and described the accident to the owner. He listened and then asked me, "Were they Mexicans, or Indians?" I said I thought they were Indians, to which he replied, "Well, don't worry, then."

I felt like a fugitive until I crossed the U.S. border. I arrived in Santa Fe just a day after David had arrived home by air. His father's health had improved, but I was a wreck.

After the Mexico experience, returning to Portsmouth was somewhat anticlimactic, but we managed to have an adventurous time there, too. I was a fairly good athlete and played on the Portsmouth baseball and football teams. One of my football teammates, Tom Van Winkle, was super bright and especially good at math. Although I was the team quarterback, sometimes in the huddle he would suggest plays. Because of his math ability, I chose him as a roommate my fourth-form year, hoping he could help me with the math homework. We became friends, although we weren't really compatible. Tom later graduated from Yale with a Ph.D. in physics. During World War II he worked with the atomic bomb research team at Los Alamos, New Mexico. After the war, sobered by his experience with nuclear weapons, he became an ordained priest and a monk and, later, the headmaster at Portsmouth Priory.

One pleasant diversion for me was sailing and motor boating from the school's boathouse out onto Narragansett Bay. Several students were experienced sailors and earned positions as skippers for the school's Herreshoff 12 sailboats. I learned enough of the basics of sailing to serve as a crewmember. My ignorance was complete, but my temptation to skipper was great.

One early spring when the bay ice had mostly melted but there were still chunks floating, I had the urge to sail, and two of us, both sailing novices, took to the sea. Never mind that the water was ice cold and sailing season was weeks away. We were out on a free day, Sunday, with no supervision. The light wind barely ruffled the calm water. We cast off and soon were quite far out from the shore.

In order to dodge a chunk of ice, I gave the tiller a sudden jerk. The boat came about, luffed and capsized, throwing us, the hapless, two-man crew into the frigid water. The bottom of the boat was now the top. Luckily, we managed to grab onto the boat's centerboard and pull our heads and shoulders above water, but the rest of our bodies dangled amid the ice cubes. The wind and tide pushed us further from the school and out into open water. Chilled to the bone, we were floating along with the capsized boat when, suddenly, our guardian angels came to our rescue: a Coast Guard floatplane flew by, spotted us, and called a cutter, which picked us up and returned us to the school. Doused with soothing sherry, we spent the night luxuriating in warm beds in the infirmary. For our punishment, we were grounded, and, needless to say, I never became a sailboat skipper.

Another maritime adventure involved the U. S. Navy. The Naval War College was located in Newport, and several Portsmouth Priory students were sons of senior officers assigned to the college. Sometime during the 1934-35 school year, a submarine, the S-51, had an accident and sank in the deep water off Block Island, near Newport. In its attempt to rescue the sub and to determine the reason for the loss, the Navy assembled a flotilla of rescue ships, complete with experienced deep-sea divers. The effort to lift the sub was unsuccessful and the mission was terminated, but part of the flotilla was still in the area and excitement still ran high, especially for students at school.

Coincidentally, at the height of this submarine disaster, my friend Goddard Kennedy, whose family lived in nearby New Bedford, was given an outboard motorboat by his father. Goddard had permission to berth the slick new boat, with its high-powered motor, at the school's dock. Goddard asked me along on a test run.

Thrilled with anticipation of a great ride, we attached the motor to the boat, pushed away from the dock and engaged the motor. The boat surged forward and off we went, zooming out into the bay. Goddard put the boat into a tight trim at full throttle and then, suddenly, the motor sound stopped. We turned around to look, and it was gone. The clamps had jarred loose and, freed from the boat, our means of propulsion sank into the deep, dark waters of Narragansett Bay. We were adrift about a half mile off shore in a motorless speedboat.

We marked the area with a buoy and rowed back to shore. That evening, one of the "Navy boys" spoke to his father, an officer involved in the submarine recovery effort, telling him about the sunken motor. Seeing an opportunity for a public relations event, the officer initiated a diver training exercise. A Navy vessel and a diving tender were ordered to the area where the motor was lost. Since it was within sight of the school, classes were recessed to watch the Navy's recovery efforts. We watched divers as they were lowered into the search area. In their big conical helmets and diving suits, they looked like human lobsters. The two ships churned the water and sent flag signals to each other. It was a highly professional mission and an expensive effort. The Navy spent two full days unsuccessfully searching for the motor. On the third day, they pulled anchor and abandoned the mission. Perhaps higher brass determined the effort would not be approved as an authorized expenditure.

Days later, some ex-rumrunners from New Bedford, hired by Mr. Kennedy, Goddard's father, recovered the motor. Running rum from Canada to the United States during Prohibition, these men had become experts at tossing contraband cargo overboard when inspectors closed in on them, and recovering it later. They would mark the spot where they dropped the goods and return when the heat was off to grapple the sunken booze and complete the illegal transaction. The task that had ended in an aborted mission for the Navy was a routine recovery for the retired New Bedford rumrunners.

In my sixth-form year, the devastating New England Hurricane of 1938 struck. I had returned to school early as a prefect to greet and orient the new boys. Their arrival and the hurricane's coincided. The storm hit Rhode Island and the school area with brutal and demolishing force during high tide, beaching ships and leaving telephone lines laced with seaweed. Bridges and buildings were damaged and over thirty-five people were killed in nearby Tiverton alone. Fortunately, our school buildings, shielded by the hill on which they were located, were spared the main brunt of the storm.

In the midst of the storm, I wrote home to my parents, telling them about my experience of the hurricane. Quotations from my letter were published in the Santa Fe New Mexican. I wrote, "The wind and rain have made a shambles of countless trees. A large pine was broken in two and fell not five feet from my window. The bay is a mass of misty, choppy waves…and we are right in the peak of the hurricane with no lights, telephone, or electric power… The entire country has been laid waste with death and destruction…"

Late that afternoon, two new boys were unaccounted for. School officials initiated a search while the storm, although somewhat abated, was still active. As day ended, the misty sky turned dark and the wind was still fierce, though not as strong.

I was assigned to search along Cory's Lane to the village of Portsmouth. This old trail traversed the boggy and overgrown area that had been the scene of a Revolutionary War battle. Tradition had it the ghosts of several Hessian soldiers, part of the British force engaged against the local colonists, had been killed in the battle and haunted the area.

In the howling wind, mist, and darkness, I proceeded down Cory's Lane, through a spooky obstacle course of fallen trees, broken branches, mud, water, and wind. I was tense and scared, my imagination hyperactive. I anticipated finding the missing boys mashed under a fallen tree or wet, lost and miserable.

As I stepped over a downed tree, my flashlight directed ahead, I detected movement and hazy outlines of what appeared to be human forms melting towards me, transparently passing through the brush and limbs. I stood frigid with fright, motionless, pointing the flashlight at the approaching figures. They floated forward, three of them abreast, in martial, drill-like cadence. Then the uniformed Hessians, with crossed bandoliers, pointed, split helmets (shaped like Bishop's miters), shouldering muskets with bayonets, passed right through me. I quit the search and returned to the school. No one believed my story.

After the storms stopped, several revolutionary war-era cannon balls were found in the area of Cory's Lane where I had searched. Perhaps, I thought, the destructive hurricane had disinterred my Hessians along with the cannon balls. And as for the "lost boys," it turned out they had missed their bus and were found in the Providence Biltmore Hotel.

During my final year at Portsmouth I focused on passing the college entrance examination. Questions on our school tests were taken from previous entrance examinations and therefore were preparatory to the real thing. The college board tests lasted four hours. I guess prayers and lucky guesses helped, since my test results warranted graduation. I hoped they would also gain me acceptance at Harvard.

Graduation day was special, even though my family saw no compelling reason to travel to New England to attend the ceremony. I didn't attend either. Instead, I spent the day on a Coast Guard cutter monitoring the J boats competing in the American Cup races off Newport. I was a guest of the cutter's skipper, a friend of my father's. The weather was perfect, with fair skies and a brisk wind whipping up a choppy, white-capped racing course. Two glamorous sailboats, one representing the United States and the other Great Britain, jousted for victory. The U.S. boat, its skipper a Vanderbilt, won the race and retained the cup for the United States.

It was a perfect graduation day. After five years at Portsmouth, I was on my way to college.

Harvard Undergraduate Days

Harvard College was not completely unfamiliar to me upon my arrival there as a freshman in the fall of 1939. My father had graduated in 1908 and remained an active alumnus and supporter all his life. Naturally he pointed his sons in the same direction. I remember his twenty-fifth reunion in 1933, and in 1936 I accompanied him to his alma mater's three-hundredth anniversary celebration, where he represented the University of New Mexico. My brother Hank, a senior in 1939, would graduate along with John F. Kennedy in the class of 1940. I had visited Hank at Harvard while I was at Portsmouth Priory School.

I was assigned to Strauss Hall, along with George MacLellan, my Portsmouth classmate and soon-to-be roommate. The first few days were swirling with decisions as we chose which sports to play, purchased our books and supplies, faced solicitation efforts from campus organizations, made our course selections, and set our schedules for classes and testing. We also began to meet some of the twelve hundred "men" in our class.

One memorable gathering during my first weeks at Harvard was dinner at John Richardson's home. Richardson's father had graduated at Harvard with my father. The other guests were Bob White, Harry O'Hare, and John Moot, whose fathers were also in the class of 1908. I also recall attending during those first few days a required reading course for slow readers. It exposed my slowness, but unfortunately did not cure it.

The freshman Union Building, the central hub of freshman social activities, contained the dining hall and meeting rooms. It was located on the opposite end of Harvard Yard and it often seemed to be just a little too far away and served breakfast too early for those of us living in Straus Hall.

My first courses were Spanish, English Literature, World History and Military Science (the Reserve Officers Training Corps, ROTC), an elective course. I chose ROTC not so much for its course content, but because it involved horses and a chance to play polo. Like Princeton, Yale, and Cornell, Harvard's ROTC provided training with horse-drawn field artillery and furnished the polo ponies for the college polo team. In 1939 horse power still gave the artillery its mobility, and equitation was an integral part of the ROTC training. I arrived as an experienced rider and was selected to be a student instructor in Hippology (horse care and riding). This gave me a chance to join the college polo team. (The horse artillery era and polo at Harvard ended in World War II, as gasoline and trucks replaced hay and horses, but I was stuck with the field artillery.) Much to my dismay, however, the ROTC course also involved considerable mathematics (geometry and trigonometry), a difficult discipline for me, which made this course one of my constant problems.

Professor Merriman taught the required world history course. I found his lectures a real delight. Merriman was an avowed snob and always dressed to the T, but his lectures were brilliantly delivered. His knowledge of the subject and his colorful presentation kept the class spellbound.

I chose Spanish to satisfy the language requirement because I was familiar with a rustic dialect of Spanish I was exposed to in New Mexico. I also believed my pronunciation, picked up among my neighbors in Santa Fe, would be helpful. But it turned out Spanish grammar and verb tenses were quite challenging for me, and after the first hour exams in November, I switched to French, which I had studied at Portsmouth.

As the fall semester unfolded, a routine developed. Classes and study occupied most of my time. The rest of my active schedule was filled with "bull sessions" with dormitory friends, participation in athletics, going to football games, and attending organized or spontaneous parties and social gatherings. Hippology on Soldiers Field near the football stadium took up a good part of my fall afternoons. I also went there to try out for the freshman football team, but this effort did not last long, as I was not big enough or fast enough to compete with the brutes that interested the coaches. During tryouts, I played on the defensive line and was befuddled because it seemed I was decked repeatedly by the same offensive man, regardless of my position. Only later did I solve the mystery of the pummeling, when I found out that identical twins played opposite me on the offensive line.

Parties in Eliot, Winthrop, and other houses occupied by upperclassmen followed Harvard's varsity football games. When Harvard played Army, it was a thrill to see the West Point Corps of Cadets march onto Soldiers Field and into the football stadium. On one occasion after a Harvard-Army game, my brother Hank hosted a cocktail party. Among the guests was a Santa Fe contingent including West Point cadet Jack Kenney and Harvard students Bergere Kenney, Stalker Reed, and Wiley Barker.

Relatives in Boston exposed me to a more sophisticated side of Boston social life. Nancy Bloomer, who was my Aunt Alice "Pat" Gross's sister, and her husband, Tom Bloomer, were proper Bostonians and comfortable members of the Social Register set. Tom had graduated from Harvard and while in college had been a member of the somewhat exclusive A.D. Club. Nancy and Tom had two attractive daughters: Nancy, who was married, and Priscilla, who was more my vintage and still single. Visiting Priscilla at their home, a townhouse on Marlboro Street, became a pleasant diversion for me.

Mrs. Bloomer saw to it that my name was included on the list of approved Harvard student invitees to the various debutante parties in Boston. Being invited to these elaborate affairs for Boston's finest young ladies required formal attire and some facility at chitchat and the rudiments of dancing. As I became more seasoned, I often wondered why a father would waste so much money entertaining a group of young men who mostly ended up getting high with each other rather than dancing with the young ladies.

The freshman year schedule followed the traditional pattern set up by previous college presidents and fine tuned by James B. Conant, the president during my time in college. The November hour examinations highlighted the fall semester. These course reviews and faculty evaluations were designed to determine how we were progressing—or regressing. The test results also helped us decide on a major and gave us an opportunity to change courses we weren't suited for. More ominously, the tests also provided information for a culling process to weed out students who were not happy or adjusting to the Harvard regimen.

Mid-term examinations came after a Christmas break, or "reading period." Winter in Cambridge lent itself to more serious academic pursuits as the weather was generally cold, chilly, and wet—a good time to be indoors, studying. With spring term came a delightful change in mood from the dark

New England winter. The ice on the Charles River floated off, allowing rowing crews to head out and slice the water with their slim racing skiffs. With the good weather, distractions multiplied. Tennis, baseball, track, bird watching, hiking, and the outdoor concerts of the Boston Pops—these made studying a test of discipline and determination.

As the academic year ended, so did my time living as a freshman in the Harvard Yard. If we passed our year-end examinations, we would move into the several upper classmen houses sprinkled along the banks of the Charles River. We had to decide on an academic major concentration for the coming year. We also had to decide where and with whom we would live for the next three years.

My freshman roommate George MacLellan and I had gotten on pretty well in Straus Hall. I can't recall what George chose as an academic major, but he too was in the Army ROTC program. George, who was rather quiet and somewhat conventional, came from New Hampshire. He was not particularly interested in social activities, although he liked baseball and politics. He had no bad habits. He earned the sobriquet "Mr. Nicely, Nicely." George never expressed how he felt about me, but we just got along in a kind of old marriage way. So we decided to continue on as roommates and chose Eliot House as our future home

During my freshman year I had become friends with a diverse set of individuals, each from a different cookie cutter. Mostly, they had attended various prep schools and had been raised in different states. My academic interests lay in history, government, and geography/geology, but being in the ROTC required that I take Military Science. I decided on a major of history and government, not suspecting military science would become my major concentration later on.

After spending a wonderful summer in New Mexico in 1940, I returned to Cambridge as an upper classman and an occupant of Eliot House, along with my roommate George MacLellan. Eliot House, an elegant building situated on the Charles River a short walking distance from Soldiers Field and the football stadium, was the most sought after of the seven undergraduate houses. The building encircled a handsomely landscaped grass patio and had its own library, a bright, well appointed dining room, and eight entries to four floors of bedroom suites. The housemaster, "Friskie" Merriman, was known for his hospitality and his somewhat snobbish criteria for selecting the house's occupants.

George and I shared one of the four suites on the fourth floor of D entry. We had our own living room with a fireplace, a shower/tub bathroom, and two bedrooms. A biddy kept the suite tidy. The house dining room was furnished with tables of various seating capacity, each set with linen napkins, silverware, and glasses. We made our meal selections from card menus and were served by uniformed waitresses three times a day.

This elegant Harvard living was soon to pass into oblivion. War was on in Europe and a drive for military preparedness was apparent as the armed forces began to flex their muscles in the United States. The mood during the academic year of 1940, however, remained pleasant and comfortable, with a few highlights to recall.

I took several American history courses, including one on the history of the South (pre- through post-Civil War), taught by Dr. Schlessinger. My tutor Barnaby Keeney later became the president of Brown University. I also took a fascinating earth history course, taught by the renowned Dr. Kirtley Mather. Military Science, now sans horses, had evolved into a mathematics course, which continued to drag down my otherwise quite respectable grade average.

I had been bounced from the varsity football squad, but I made the Eliot House team. The climax of the season came when we played at Yale against Jonathan Edwards College, our sister residential hall. This entailed playing in New Haven and a post-game bash in New York City.

One tragedy marred the year: the Coconut Grove fire in downtown Boston. The fire broke out in the popular nightclub just when the place was packed with boisterous football fans celebrating a win by Holy Cross College. I had been in the club in the early evening, but, fortunately, I left before the disaster struck. My friend and Harvard classmate from Santa Fe, Nathan Greer, however, was caught in the center of this fiery cauldron. With his date grabbing his wrist, he made for the revolving exit door, only to find it jammed by frantic escapees. Nate managed to pull the blocking key, permitting the door to revolve, but his date lost her grip on his wrist and was trampled to death by the frantic stampede. She was among the one hundred-plus who died. When her body was recovered, she still clutched Nathan's elastic-band wristwatch in her dead hand. Nathan survived but was scorched, physically and mentally.

During the summer of 1941, two classmates visited Santa Fe as my guests. The first, William Clyde Josiah Sharpe—known irreverently as Buck—was a boarding school orphan, having been sent away to school when only about nine years old. He graduated from Choate School in Connecticut and I met him early our freshman year at Harvard. Buck was a natural football player, stocky and strong; he probably weighed about two hundred pounds. He kept his blond hair cropped short. His natural shyness melted with familiarity and was replaced with a gentle but astute sense of humor. Other than his boarding school "home," he claimed Minden, Nevada, and Swansboro, North Carolina, as his places of origin, but these were little more than infrequently occupied havens owned by his much-married father, Dr. Sharpe.

Dr. Sharpe and Franklin Roosevelt were both in the Harvard class of 1904. Sharpe had grown up in Pennsylvania and came to Harvard on a rowing scholarship. His ambition was to be a medical doctor and, other than rowing, he studied and worked at part-time jobs to supplement his otherwise meager income. One of the required freshman courses in his time was Physiology I, a basic study of the human body. Sharpe excelled in this course and kept detailed notes. He was thoroughly prepared for the November hour exams on the subject. Just two nights before the scheduled exam, he was in his room studying when a knock sounded on his door. He admitted Franklin Roosevelt, whom he did not know. Roosevelt made a brief and somewhat imperious request to borrow Sharpe's course notes, because Roosevelt had missed a number of class sessions. Sharpe obliged him and loaned the notes, but the next day Roosevelt returned, complaining he could not decipher Sharpe's handwriting. Sharpe took on the task of preparing this future president for the exam. As a result Roosevelt earned a passing grade and Sharpe earned a lifetime of appreciation. Sharpe became a regular tutor for Roosevelt and his coterie of Gold Coast friends. The relationship blossomed into a much-needed financial windfall for Sharpe. Later in life, President Roosevelt nominated Sharpe to be the U.S. Ambassador to Brazil.

Dr. Sharpe went on from Harvard to become a well-known brain surgeon and was active in medical missionary work in South America. His professional career left little time for parenting Buck. When he was not deeply involved in his work, he would retreat to his remote but palatial hideaway in Nevada, or to his six-thousand-acre plantation in rural North Carolina.

Thus, my friend Buck Sharpe really had no family life, and when he was not in school, he was under the direction of a guardian. When visiting the Kelly tribe in Santa Fe, he enjoyed being included as part of the local gang of friends and family, who kidded, loved and entertained him in an environment entirely different from any he had known. He flourished when he was with us.

In turn, I was his guest on his father's remote coastal plantation in North Carolina during the spring break.

Buck served as a Marine fighter pilot for five years in the Pacific during the war. Perhaps influenced by his brilliant but somewhat bizarre father, he later became a physician with a specialty in psychiatry.

The other Harvard classmate to visit me in Santa Fe that summer was Joseph P. Downer, an army brat whose father, a Virginian, had graduated from the Virginia Military Academy as a regular army officer. Colonel Downer fought in the Philippine insurrection under General Arthur MacArthur, military governor of the Philippines. In World War I, he commanded the Sixteenth Field Artillery Battalion, which fired the first U.S. salvo at the Germans. An avid horseman, Colonel Downer was a member of the U.S. equestrian team at the Olympic Games of 1921. In the 1930s, while his father was stationed at Governors Island in New York Harbor, Joe attended Trinity School in New York City.

I met Joe in ROTC. Raised around soldiers most of his life, he stood out in the group of raw recruits who had just joined the Military Science unit. He was a conscientious student, somewhat straight laced, and he excelled at math. His ambition was to do well at Harvard. He came to my rescue by helping me grasp the geometry and trigonometry involved in computing artillery problems. From this interaction, a strong friendship developed, and in my junior and senior years we were roommates. By then, his parents lived in Bellport, Rhode Island, near his father's post at Camp Upton.

During vacations, I became a frequent visitor at the Downer house. I never knew anyone quite like his parents. The colonel was a disciplined and demanding officer, but his wife, Gladys, definitely outranked him. The two would frequently battle at the dinner table, which I found amusing even though the arguments sometimes became quite heated. I would take sides depending on the nature of the topic they were arguing about, although I tried to be diplomatic about expressing my opinions.

Because of Colonel Downer's prominent position as the commander of the main army reception station for New York City, he was frequently invited to athletic events, city dinners, and opening-night affairs in art and theater circles. When we were included, these events gave Joe and me access to places, people and happenings that otherwise would have been off-limits to two college boys.

When Joe visited me in New Mexico, we hiked a lot. One trip took us on a walk from the vicinity of Mora to Santa Fe. We ended this trek in Nambé at a party hosted by Nancy Thompson, a Radcliff/Harvard student and frequent visitor at our digs in Cambridge. My sister Yei was at the party, and when Joe met her, his interest in New Mexico became much more focused—on her. His attraction to my sister went unrewarded but later he introduced her to his army friend Frank Gorham, whom she ended up marrying. I take credit for the romance, since it started, indirectly, with me bringing Joe to Santa Fe. Downer became a lifelong friend and a frequent visitor to Santa Fe. His ambition to do well at Harvard came to fruition later, when he was elected president of the Harvard Alumni Association.

My junior year was one of change. Joe Downer and I moved into a larger suite in the D entry of Eliot House. This suite of rooms included a living room that was the center of activity. There were twelve of us in the suite. We comprised a homogenous group, representative of the student body, with interests in our studies as well as athletics. We enjoyed life. Those of us living on the fourth floor of Eliot House became known as the Roaring Forties.

Another change that year involved social clubs that accepted as members only a minority of the student population. Membership in such a "final club" quite often stemmed from being related to a former club member. In addition, class office positions and involvement with academic and athletic activities were helpful for gaining club membership.

Four of my roommates were members of the A.D. Club, which was founded in 1836. It and the Porcellian Club were considered the leading college clubs. Downer and I were somewhat outside the loop, but our roommates were good politicians and helped us gain admission to the A.D. in our junior year. The friendship, facilities, and convenience of being a club member were more important to me than the social status that came with membership. (Now, when I visit the college as an old graduate, the A.D. Club seems like a home away from home.)

Clubs by their very nature are endangered species. The Harvard administration now frowns upon the so-called elitism and the quasi-independence of the clubs and has been active in its efforts to eliminate them. In 1941 and 1942, however, club life was pleasant, active, and healthy.

What was not healthy was the war in Europe and the rise of Japanese aggression in Asia. As Churchill charmed Roosevelt, France fell, and British forces were dislodged from the continent, pressure for U.S. rearmament and greater army manpower increased. The draft was on and college deferments became the exception rather than the rule. Obviously, the ROTC programs became more popular and important.

December 7, 1941, the Day of Infamy, when the Japanese bombed Pearl Harbor, initiated the unraveling of the social order at Harvard and the splintering of my class. Despite an appeal from Harvard President Conant to students, asking them not to overreact, enlistments and draft calls disrupted the college routine and the waiting game for service call took over all our attention.

In early 1942, when I was home on break, my mother and I attended the Snake Dance on Walpi mesa. The dance stopped abruptly as the tribal leaders looked down from the rooftops on the immobile dancers and awestruck visitors. The silence was broken when the pueblo governor announced the Hopi Nation was declaring war on Nazi Germany. What a dramatic moment on a mesa top in mystic Hopi land.

In January 1943 half of my ROTC class members, both Army and Navy, were called to active duty. My time came in June. There was no graduation ceremony; we received our degrees along with reporting orders, and college life was over.

Bud and Hank Wearing Their Father's WWI Uniform in Front of Federal Building, Santa Fe, ca. 1923

Bud, Caroline, and Hank, "Lincoln Avenue Indians," ca. 1924

Bud and Hank, "Cowboys" in Las Vegas, N.M., ca. 1926

H. W. Kelly Residence, Sixth Street, Las Vegas, N.M.

Grandfather H. W. "Harry" Kelly, n.d.

Portsmouth Priory School Cub Baseball Club, ca. 1936

Harvard Freshman Yearbook Photo, 1939

Harvard Freshman Polo Squad, ca. 1939

Bud with Dave Davenport, Xochimilco, Mexico, 1940

Father and Son, Two Daniel Kellys, ca. 1942

1943-1945

WORLD WAR II

I reported for active duty in the Army on June 10, 1943, after being in the Harvard ROTC unit since the fall of 1939. Instead of receiving an officer's commission, I was designated a buck private, like everyone else at the reception center at Ft. Bliss, Texas.

Nate Greer and I flew down to El Paso as ordered. We were stripped of our Brooks Brothers suits and, donning government issued, poorly fitting uniforms, we marched into the steaming cauldron of sweating raw recruits. We were issued sheets and a blanket, assigned to a barracks, and instructed on how to make our canvas cots into beds. The sergeant in charge of our provisional company then assigned us duties. Noting we were Harvard graduates and therefore capable of discharging responsible assignments, he made me the latrine orderly and Greer responsible for picking up cigarette butts and trash.

Willie Palou, who prior to his drafting had been the long popular leader of Santa Fe's La Fonda Hotel orchestra, led the reception center's marching band. By this time, Willie had attained the rank of senior sergeant. When he spotted us, he pulled us from our formation and greeted us warmly. His group of musicians, all senior soldiers, invited us to join them on a night out in Juárez, Mexico that very night. Thus, within twenty-four hours of joining the army, we were absent without leave (AWOL). Shielded by Willie's group of seasoned reception-center NCOs, we joined in a boisterous evening in Juárez. No harm came to us as a result of this escapade, but the latrine and cigarette butt assignments continued.

Fortunately, ten days later, on June 21, Greer and I were ordered to a Military Police battalion in Springfield, Connecticut. I was the troop commander for the journey, the "troop" consisting of the two of us. Maintaining discipline was difficult. On the train, we met Tommy Mansville's just-divorced, fifth trophy bride, who invited us to be her guests at her New York apartment. Despite this alluring prospect, when the train arrived at Springfield, our "detachment" followed orders and reported to the Eighth Prisoner Escort Military Police Battalion.

The battalion was made up of older men, classified fit for limited physical duty. They ranged in age up to the early thirties. Quite a number of them were Jewish. Ironically, the ultimate mission of this unit was to escort German Africa Corps prisoners back to the United States. We received basic infantry training: hiking, handling small arms, and work details. The facilities afforded this unit were deplorable. The barracks was formerly the exhibit hall for a county fair. Toilet facilities offered two heads and three shower stalls. Our captain was probably overage in grade (too old for his rank). His drilling exercises emphasized marching, overnight bivouacking, and stealthily approaching terrain features. We coined his training regime "creeping and crawling with iron balls."

Greer and I were ordered to this unit intentionally, because of our ROTC background. It was a stopgap measure pending our scheduled assignment to officer candidate training school at Ft. Sill, Oklahoma.

While in Springfield, probably because of the limited hygiene facilities, I developed a severe case of folliculitis, an inflammation of hair follicles. I erupted with boils, and after a series of boil-squeezing treatments failed, I was assigned to the Springfield military hospital.

My initial treatment was to be immersion in a bathtub of treated waters, to eliminate the surface infection. My turn in the tub followed a patient who had an open ulcer. When I asked him what his problem was, he responded, "Shancroids." I was appalled, as I knew something about this sexually transmitted infection and how it could be spread, other than the usual way. Despite my lowly rank, I insisted on seeing the medical officer in charge of my ward. Fortunately, he listened to me and transferred me out of the venereal disease ward to which I had mistakenly been assigned.

Although I was being treated for a dermatology disorder, I was bunked in a psychology ward whose occupants were Section Eight patients, soldiers who had gone bonkers or who pretended to be so afflicted. This open ward was a zoo. Some men thought they were roosters and crowed constantly. Others perched on the edge of their iron cots, poised for flight and gibbering in some tongue only known to their confused minds. Passing the nights there was like being in a monkey cage, with jungle noises and acrobatics making sleep impossible.

My condition quickly improved, probably expedited by a burning desire to get out of that chaotic environment. Although not entirely recovered, I returned to the battalion and received orders to report to the field artillery officer training course at Fort Sill, Oklahoma.

Fort Sill, near Lawton, was the headquarters and training area for the field artillery branch of the army. Located in the heart of Indian country, it was founded in the mid-1850s. The need for space for working on maneuvers and firing at distant targets made it one of the largest army posts in the United States.

I was assigned to Officer Candidate Class 88 in July 1943, in the middle of the roasting summer heat. Our class of candidates was housed in small, four-man frame hutments in an area with no shade or shelter from the prevailing winds. Responsibility for keeping the housing unit spotless rotated weekly among the hut mates. We ennobled the duty with the title Hut Slut.

We were trained under extremely high-pressure conditions. They piled on the courses—Gunnery, Communications, Materiel, and Tactics—in purge doses, as we had only thirteen weeks to complete the entire training regimen. Knowing failure to complete the training would result in assignment to the infantry as lowly privates goaded us on toward satisfactory completion and an officer's commission. Our only respite from the grueling pressure came on Sundays, when we would descend on officer candidate David Davenport's tiny rental house in Lawton. There his wife, Effie, fed us her home cooking and the fresh look and scent of femininity.

On November 23, 1943, I received my commission as a second lieutenant of field artillery, with orders to report to A Battery of the 871st Field Artillery Battalion supporting the 263rd Infantry Regiment, 66th Infantry Division, located at Camp Joseph T. Robinson, Little Rock, Arkansas.

I was somewhat annoyed with my father, who was responsible for my assignment. As was customary with him, he felt his personal touch would help my military career. His cousin Colonel Clarkson McNary, a West Pointer, commanded the 263rd Infantry Regiment. My father suggested to Colonel McNary that I be assigned to his command. As a result, I was the only member of my Officer Candidate School class directly assigned to a combat division; the others were ordered to more advanced, pre-assignment training courses. I reported to the 871st Field Artillery Battalion and spent the next six weeks in the snowy, muddy, freezing hills of Arkansas on division field maneuvers.

I was the junior officer in A Battery, under the command of Captain Pritchard, a laid back, easy going Virginian. As the only second lieutenant, I was assigned to a variety of duties none of the other officers wanted. Among my responsibilities was conducting short arm inspections, a rather up close and personal inspection for evidence of venereal disease, on payday. I also organized dances—with the requirement that I locate willing and attractive females as dancing partners for the enlisted men—and served as mess officer and motor officer.

One of the training activities at Camp Robinson was running the infantry units through maneuvers while they were exposed to live fire from machine guns and explosive devices. The training took place in a large, isolated area equivalent in size to several football fields. To maximize the realism of this training exercise, the area was blistered with shell craters and crisscrossed with barbed wire entanglements and "enemy" defensive positions.

I supervised this training drill, assisted by two sergeants who were highly qualified and experienced in handling explosive missiles and were permanently assigned to orchestrate the exercise. My responsibility was to brief the soldiers involved in the exercises and launch them into action.

One day things went wrong with the live-fire drill. The simulation itself went well. The realism was intense as soldiers crawled along under machine gun fire, dropping into booby-trapped craters and, ultimately, advancing and overrunning the mock enemy positions. At the end of the day, after running several hundred men safely through the drill, the two sergeants and I were evaluating the day's work. The sergeants were repacking the explosive devices into containers to make ready for the next day's training. As I turned my back to the men, an explosion ripped through the quiet. I turned around to see all that remained of the two sergeants: gruesome body parts strewn on the ground. Apparently the "experts" had somehow triggered the devices they were handling and had blown themselves to bits.

I was appalled by the sight, and in shock. I remember picking up the parts and driving the jeep to division headquarters to report the accident. I cannot recall any of the details of the investigation or follow-up on the tragedy. Luck was on my side that day.

Shortly after the winter maneuvers, I was sent back to Fort Sill to attend the survey school. It was essential to the effective use of artillery to accurately map the location of gun batteries in relation to each other and to reference points in the target area. This task required detailed mapping and drafting of firing charts. Survey sections organic to field artillery units performed this task, which obviously demanded a thorough knowledge of geometry and trigonometry.

Mathematics was anathema to me and I had avoided it like the plague in school. My aversion to math made the gunnery course, which was basic to artillery training, mental torture, but somehow I survived. I then was picked to be trained as the battalion survey officer in preparation for the critical D Series maneuvers, which would determine whether the division was combat ready.

I reported to Fort Sill again to spend the next two months in agony as I tried to master artillery survey. Occasional dates arranged by my old friend Lieutenant Nathan Greer, who was an instructor at the Cooks & Bakers School at Fort Sill, served as a critical diversion from my inferno.

I completed the course. Upon returning to the division, I was assigned to prepare the fire control survey required by our battalion in its role in the D Series tests. The artillery's mission was to furnish accurate supporting fire for our attacking infantry regiment. This required placing a wall of bursting shells ahead of and above the advancing regiment. Calculations were critical; if the projectiles exploded short, they would decimate our troops, and if the rounds burst too low or too high, they would be ineffective against the enemy.

The deadly, live-ammunition exercise required the most accurate survey control. The survey crew, of which I supposedly had command, supplied this critical information. We spent two days mapping in the field in advance of the exercise. Fortunately, all of my enlisted crew members were professionals, so we completed the survey control and integrated it into the exercises. The test's objective was to dislodge "enemy forces" occupying a low lying ridge to the front of our regiment. Artillery fire on the defended ridgeline initiated the attack, which was followed by protective barrages in advance of our attacking troops. The artillery support proved to be accurate and effective. We took the ridge and suffered no "real" casualties as a result of the live artillery support. The survey crew had done a superior job and I reflectively shared the credit for their achievement.

The division did well on the D Series test, earning a "combat ready" rating. Our overseas assignment was pending. Troop replacement needs in both the Pacific theater and in Europe, however, drained the division of large numbers of its trained men, who left to replace losses suffered by combat units. As a result, we had to train and integrate green replacements into our infantry regiments, which delayed our shipment overseas.

In May 1944, A Battery was detached from the division and moved to Fort McClellan, Alabama. Our assignment was to provide artillery support in the training of the 442nd Japanese Nisei Infantry Regiment—a regiment that was later assigned to the Eighth Army in Italy and became one of the most highly decorated units in the army.

While we enjoyed a pleasant escape from the hot weather to spend the summer at Fort McClellan, the division transferred to Camp Rucker, Alabama, a hot pesthole near the Florida border. Our battery was stationed in the hills overlooking Georgia. Our unit savored the independence from the division, and the troops in training were efficient and responsive. As an added bonus, Atlanta, a friendly city, offered us a pleasant and nearby escape from our isolated military outpost.

Our battery rejoined the division at Camp Rucker in mid-summer of 1944. I was sent to the Chemical Warfare School at Edgewood Arsenal, Maryland, to learn about defense against poisonous gas attacks. I returned there in September 1944 and resumed my duties as battery assistant executive, motor officer, and now, chemical warfare officer—and anything else they thought to assign me.

In a totally surprise move, on October 25, 1944, I was transferred from A Battery, 871st Field Artillery Battalion, to Headquarters Battery, 66th Division Artillery, and assigned as aide de camp to the division artillery commander, Brigadier General Francis W. Rollins.

General Rollins, a New Englander from Providence, Rhode Island, was about fifty years old, handsome, slightly built, mustached, well groomed and reserved—indeed an officer and a gentleman. In civilian life he had worked as a bank executive. A veteran of World War I, he had maintained his involvement in the military as a National Guard artillery officer and was federalized when World War II was declared. In April 1943 the 66th Division was activated and Rollins was appointed the division artillery commander.

Rollins's deputy commander, Colonel R. Dinwiddie Groves, also served in World War I. He was a colorful political officer who had been in the Missouri National Guard. Somewhat overage in grade and a friend of Vice President Harry Truman, he was placed on active duty and assigned to the 66th Division. Although he was about sixty years old, he and I became good friends.

Responding to the needs of these two senior officers became my new responsibility. An aide de camp essentially acts as a gopher for a general, a loyal private secretary, basically answerable only to the general and, to a lesser degree, responsible for the general's well being. An aide is supposed to possess social graces and a military bearing and be willing to take on any assignment the general gives him. I preferred being a battery officer, but serving General Rollins did offer a variety of duties more interesting than the routine duties of a troop command.

In early November our division was ordered to the New York port of embarkation for shipment to Europe. Moving several thousand men and equipment from Alabama to New York City and stuffing the units onto a troop transport proved to be a demanding task.

As an aide de camp with no actual troop responsibility, I was in charge of our advance-party detail, with particular responsibility for shipboard arrangements for the general and division artillery headquarters personnel. Our ship, *RMS Britannic,* a converted Cunard passenger liner, was docked on the 42nd or 43rd Street pier. This behemoth of a ship would carry some fifteen-thousand-plus soldiers and the necessary cargo to feed and bunk them across the Atlantic.

Tending to immediate housekeeping details for division artillery headquarters, I found where the senior brass was to be billeted. As I recall, the general was assigned a compartment containing a single bunk and an adjoining head in an area that was probably a dressing room in the days of private passenger use. Several officers and I were assigned to a "stateroom" consisting of three tiers of bunks opposite each other with just enough aisle room to get in and out of them. The troop compartments in the hold of the ship consisted of row after row of steel-framed, canvas bunks stacked five high in a cavernous, poorly ventilated, steel-decked area that was to be the troop barracks for the ocean crossing. These quarters reminded me of a huge can of sardines.

One bright spot was the loading of foodstuffs destined, I thought, for the meals to be served en route. There were slabs of frozen beef and sheep carcasses being stacked in refrigerated storerooms, and crates of eggs, frozen vegetables, and cases of wine and bottled beverages filling the cargo area. As it turned out, these mouth-watering supplies were intended for British civilian consumption and not for the pleasure of the canon fodder crammed in the troop ship.

Once the troops boarded and settled in their bunking areas, loudspeakers called all troops on deck for briefing. With so much humanity in such a small space, all activities—messing, calisthenics, abandon-ship drills, etc.—were jammed into tight schedules. The ship offered no space to just "hang out." When not otherwise engaged, the troops pretty much lay on their bunks. They were served meals twice a day in chow lines; the food was placed in each man's mess kit and he consumed it while standing up. Once the program became routine, confusion ceased and the disciplined activities became routine.

The *Britannic* was a British ship, manned by Brits, and the food served confirmed that country's sorry reputation for cooking. Seasickness was prevalent among the troops. Being packed in a poorly ventilated bunk area, rocking in unison with the rough Atlantic, was a sure formula for *mal de mer.* The frequent serving of Sea Pie Nelson, a traditional and favorite British dish, compounded the gastrointestinal distress. This concoction, slopped unceremoniously into mess kits, contained cold hunks of fish and bread swimming in a watery broth, garnished with floating fish scales.

U.S. Navy destroyers escorted the *Britannic* and a long convoy of vessels headed for Europe. The threat of submarine attack was ever present. We participated in abandon-ship drills several times each day. We had no idea when to expect the drills. When the sirens sounded, all units rushed to their assigned deck areas, not knowing whether we were engaged in a drill or the real thing. Our convoy sailed an erratic course to (hopefully) bypass areas frequented by German subs. We were at sea for over two weeks, sailing south to the coast of Africa and then north to our destination, Southampton, a major port on the English Channel. We landed there on December 8, 1944.

Our division disembarked and proceeded to an assembly area near Dorchester in Dorset, about fifty miles from Southampton. Division Artillery Headquarters occupied an unfurnished, unheated, three-story brick building that had been worn out by nearly continuous troop occupation since the start of the war. A single coal-burning fireplace heated each floor. Our fireplace collapsed and the chimney fell down shortly after our arrival.

We kept busy in Dorset, matching trucks and cannons with the firing batteries, drawing equipment from depots located in various places in southern

England. It was cold, and daylight only lasted a few hours before darkness set in. Our time there gave us a not-so-pleasant initiation for even less pleasant circumstances we were soon to face.

The war in France was going badly. The Germans had mounted a massive, surprise offensive, later named the Battle of the Bulge. They aimed to penetrate the line of allied forces by attacking through the Ardennes Forest on the French-Belgian border. They cleverly planned to make their move at a time when poor weather prevented allied aircraft from flying. U.S. forces desperately needed reinforcements, which is why our division was ordered on Christmas Eve to cross the English Channel and proceed to Bastogne.

The day of departure happened to be payday, but there was no time to disburse the funds. We had little need for the money in any case, since all units were loading onto ships for transit. General Rollins ordered me to personally safeguard the troops' money until the unit arrived in France. There I would return it to the ordinance section for distribution, as the pay had been counted out in British pounds and we would be in France. I can't recall how much money it was, but it was a sackful. I think I ultimately placed it in the ship's safe on the LST (Landing Ship Tank).

So, on Christmas Eve we found ourselves crossing the English Channel en route to Cherbourg, France, with infantry regiments on troop transports and artillery battalions aboard LSTs crewed by the U.S. Navy. Loading was not tactical.

Two of the three infantry regiments, plus some support units, sailed on his majesty's transport, the *Leopoldville*. I was on an LST some distance ahead of the transport. My memory of events on our crossing is blurred. I had developed conjunctivitis and by the time I boarded the LST, my eyelids were almost totally glued together. I remember the division artillery surgeon leaning over me while speaking to a navy corpsman. At the conclusion of the conversation, the medic opened the ship's sealed liquor compartment and handed the doc a bottle of rum. The doc gave me a big swig and probably favored himself the same way. With that, the ship's sacred liquor store was breached and I sacked out on the ship's deck.

At dusk, about six miles off Cherbourg, a torpedo struck the troop ship *Leopoldville* amidships, directly into the troop compartments. In a matter of minutes, these highly trained combat units were reduced to a mass of dead and wounded men bobbing and freezing in the choppy waters off the coast of France.

It was almost dark when our LST beached. The torpedoing and resulting chaos disrupted what would have been a smooth debarkation. After a long cold wait, our vehicles, with their headlights glowing like green cat eyes as the only illumination, started rolling inland. I sat in the back of a three-quarter-ton weapons carrier, the last vehicle in our convoy, still barely able to see. The commander of our section of the convoy (our "serial") was Warrant Officer Crosby, regular army, from the motor section of Headquarters Battery. I knew him slightly, but was never very crazy about him.

Around midnight we arrived at St. Lo. Because of the town's strategic location, it had been the scene of furious fighting. The ultimate U.S. breakthrough there had triggered the sweeping allied advance into the heart of France. Our convoy halted in the town's center, which was in rubble, although the roads had been cleared. From this hub, direction signs pointed every which direction. As I huddled in my truck, squinting at the scene of destruction, Warrant Officer Crosby called out to me. He was bewildered and didn't know which sign to follow. I scrambled out of the vehicle to talk to him. Holding a shaking flashlight and map, he informed me that since I was the ranking commissioned officer (a second lieutenant) in that serial of the convoy, I was now in command of the motor column. So much for rank having its privileges, especially for a blind-eyed second lieutenant.

The glue in my eyes seemed to vanish immediately as I took the map and flashlight and consulted the confusion of signs. Fortunately, I picked the correct road and we eventually caught up with the main column and arrived in Rennes, where the remnants of the division were reorganized for its mission in Brittany.

Because of the tragic losses on the transport, the crippled division proceeded to the Brittany coast, instead of assembling near Cherbourg to proceed eastward toward the front near Bastogne. Our new mission was to contain the Germans holding the key Atlantic ports of Lorient and St. Nazaire. Our division commander, General Kramer, was designated the Allied Commander of the Twelfth Army Group Coastal Sector, whose headquarters lay in Rennes.

Our division artillery took up positions in support of the mostly French forces encircling some thirty-thousand Germans entrenched in defense of the submarine pens in Lorient. General Rollins became the commanding general of Lorient West Coastal Sector, with a few of us making up his incomplete sector staff. The exalted headquarters of the sector consisted of a cluster of moldy dugouts recently vacated by the Ninety-fourth division. I shared one with Captain Brown, the survey officer. He was very competent and soon had our firing batteries accurately surveyed and tied into communications with the fire direction center.

Brown had his moody side. The static nature of a siege warfare mission soon bored him. Our shared dugout, a sandbagged depression covered with tin roofing, contained two sleeping *bancos* parallel and opposite each other. Rats—big, Brittany rats—roamed the dugouts, becoming especially bold at night.

Brown, a fitful sleeper, would often sit on his bunk, flashlight at the ready, sipping calvados. He was a good shot and kept his Colt pistol handy. One night when the rats were particularly active, Brown declared war. He placed a smooth, free rotating broomstick between the bunks, and underneath it he placed a tub of water. Checking his pistol, he took a big swig of calvados and waited. I lay in my sack directly across from the other end of the broom handle, directly in the line of fire. As the night dragged on and no rats ventured forth, I fell asleep. Brown, sitting on his bunk, dozed off, too. Suddenly a noise startled him. He readied his pistol, snapped on the flashlight, and lo! it illuminated a rat thrashing in the tub below. Animated by the calvados, George fired in the rat's general direction, scaring me to death. When I screamed at him, he came to his senses and stopped shooting. He had missed both the rat and me, but could finally claim a victory over a drowned rat. I dissociated myself from Brown after that and found another nesting place free of him—but still populated by numerous rats.

In our entrenched situation, the value of infantry patrolling was questionable. The area was heavily mined and thickly wooded. Thus, ours was an artillery war, where aerial observation against an ensnared enemy gave us a substantial advantage. Our artillery aerial observation section kept under constant surveillance (German troop movements) as well as activity around the submarine pens.

The observation unit flew L-4 Piper Cubs—single engine, fixed gear, canvas-clad airplanes that could land on short runways and had simple maintenance requirements. Major Pinkney, the commander of the detachment, was a genial Virginian who, when not flying, enjoyed rum and, when available, Scotch whisky or calvados. We called him Pinchbottle. I flew spotter missions whenever I had the opportunity, usually teamed up with Pinkney as the pilot.

One day during a spotter mission we had a foul-up. The scheduled plane developed engine trouble and the pilot had to abort his patrol. Since

Pinkney and I were not on call, we were playing cards and sampling calvados when the call came for us to substitute for the grounded team.

Invigorated by the calvados, we cavalierly waved off the crew chief from his usual preflight check and took off. We had been in the air about ten minutes and were flying over no-man's land, a patchwork of abandoned, tree-lined farm fields, when, suddenly, the engine began to sputter. Pinkney sobered up quickly and saw that the outside fuel pump cap on the plane's nose had not been removed, making the float valve inoperative and starving the engine of fuel. He pointed to a long rectangular field ahead of us, crabbed in, and landed. I jumped out, unscrewed the cap, and slithered back into the plane.

Pinkney hit the throttle and the plane lurched back into the air. We flew beneath an abandoned telephone line, clipped the tops of poplar trees, and headed back for friendly territory. I found myself thanking my lucky stars once again. When we returned to the airstrip, thoroughly chastened, we made a pledge: calvados was verboten and we never again ignored the crew chief.

The Lorient pocket was an area along the coast about forty miles wide and, depending on the shape of the coastline, from a few hundred yards to fifteen miles wide. Most of the allied units holding this pocket were recently-surfaced French troops. There were French Marines, a grounded air force squadron, and remnants of North African Spahi regiments (light cavalry regiments of the French army, recruited primarily from the indigenous populations of Algeria, Tunisia, and Morocco), as well as Maqui and guerilla groups. It was critical for us to identify these organizations, determine their size, combat capability, leadership, logistical needs—and their legitimacy.

The French military could not really vouch for a unit that called itself Le Dix Septieme Battalion du Marche. It was drawing rations, ammunition, and other materials from our headquarters via French army channels, and it occupied an advance position, a salient very near the main German line. Higher headquarters requested that General Rollins furnish information on the unit. The general ordered me to find the unit, assess its strength and capability, and determine its legitimacy.

Not confident of my poor French, I took along with me Corporal Hebert, a New Hampshire-born, French-speaking Canuck. As we worked our way toward the German lines through a maze of trenches, dugouts, and concealed paths manned by French troops, I imagined this was what World War I battlefields must have been like. We asked everyone we encountered about this mysterious Seventeenth Battalion. The reply was a wave of the arm, generally pointing the way. As we proceeded, the trenches became more makeshift and rundown, with fewer sandbags and more battlefield debris.

Rounding a bend in the line, I came upon several men, each wearing a grubby black beret tilted over one eye. Their wore *sabots* (wooden shoes), and bandoliers hung from their shoulders. They were pretty spooky looking. They ordered me to halt, and I knew they meant business. I rallied my courage and in my not-so-fluent French asked if they knew of the Seventeenth Battalion. They nodded arrogantly and stated that they, indeed, were it.

I was astounded to learn later that the Dix Septieme Battalion du Marche was made up of Spanish Republican soldiers who had escaped to France and gone underground as guerillas. When the allies liberated the area, these fighters surfaced and became, in their own eyes at least, a legitimate unit of the French army.

I identified myself. It was obvious I was not French, but they liked me, probably because I was a U.S. soldier, a liberator by any other name. Struggling to communicate, I mumbled out some New Mexican Spanish—*amigo, como está usted, como se llama*—and this broke the ice. They opened the calvados bottle and we soon were bonded.

The men were thirty-five to forty years old. They looked tough, and most of them were half drunk. They hated the Germans. They were full of bravado and boasted they would beat the Krauts by exposing themselves to fire and drawing them out. Because they did draw fire quite brazenly, none of the other units wanted to be near them.

I reported all this back to the general, who passed the information along up the chain of command. My reports satisfied higher headquarters and the Dix Septieme Battalion du Marche became officially legitimate.

One of my most dramatic memories is the final, formal surrender of the German forces facing us at Lorient. Attrition, weariness, our constant shelling, and no possibility of escape made surrender inevitable. Negotiations were held under a flag of truce and on May 8, 1945, in a field near Caudan, it ended. The Germans, ever disciplined, were marched as units to a weapon drop. They stripped themselves of their armament and helmets and then proceeded barehanded to assembly points.

The German commander, Lieutenant General Wilhelm Fahrmbacher, a spit-and-polish career officer, saluted General, Kramer, handed him his holstered Luger and was escorted off the field. The French army took over the details of handling the German POWs.

Within days of the surrender, our division was ordered to Germany. I remember crossing the Rhine on a pontoon bridge at Cologne, the city a vast skeleton of roofless, bombed-out buildings. The only undamaged structure was the domed cathedral. We set up headquarters in Sobernheim. The German army surrendered shortly thereafter.

I was involved in the liberation of a displaced-persons prison, where we found starved inmates, gaunt automatons dressed in striped, pajama-like clothing. Their subhuman appearance horrified me and I remember being fearful of contacting some form of contagion from these pathetic prisoners. Shortly after the surrender, the Sixty-sixth Division was ordered to proceed to Marseille, France, presumably for redeployment to Japan.

I was not involved in the division's move. Colonel Groves, our assistant division artillery commander, wanted to visit the battle sites where he had served during World War I. He invited me to join him for this sentimental journey. General Rollins approved.

Colonel Groves was the spitting image of what an army colonel should look like: tall, trim, with gray hair and a full mustache. He was an imposing military presence. But in contrast to this somewhat intimidating appearance, he was friendly, humorous, a good storyteller, and generally a pleasant individual to be with.

World War I wounds from four years of trench warfare and scarring from continual bombardment were still apparent in the area in France we visited. Withered grass and anemic looking trees covered a pot-marked, rolling country still recovering from the devastation inflicted on it thirty years before.

We traveled well together and had a good time. Our driver, a corporal, spoke Quebecois French, my school French came in handy, and the colonel's amusing recollections of being a doughboy warmed the friendly people along our route. We visited Verdun, St. Mihiel, Belleau Woods, the Argonne area and other places where the American Expeditionary Force of 1918 had fought. Our journey took us down almost the length of France, ending north of Marseilles at Salon, the division artillery's new headquarters.

August 14, 1945, was V-J Day—Victory Over Japan Day—the date of the Japanese surrender that officially ended World War II. All the U.S. troops jammed around Marseille, waiting for redeployment to Asia, greeted uproariously the news of Truman's bombing of Hiroshima and Nagasaki. The exuberant show of rockets and tracer bullets, accompanied by wailing sirens, made a U.S. Fourth of July celebration look like a tea party.

Now that the party was over, everyone wanted to go home, but moving more than a million men from Europe to the United States was a horrendous job. The mouth of the funnel was wide, but only a small stream could flow through at one time. Top priority was given to the men who had served overseas the longest; that left a considerable challenge in trying to figure out what to do with those who had to wait. Part of my duties involved one approach to this problem.

The army arranged for colleges and universities in France and England to open their doors to qualified soldiers, primarily men who had enlisted while in college or who had sufficient pre-college credits to warrant temporary enrollment in these participating universities. Quotas were allocated to the units waiting for redeployment. I had the job of administering this program for our unit. In Britain, slots were available at Oxford, Cambridge, Aberdeen, Glasgow, the London School of Economics, and others. During the summer of 1945, I filled all my quotas for the fall semester. In doing so, it occurred to me that, since I probably had to wait another year before I would be eligible to return to the United States, maybe a few weeks at a university would be an interesting way to pass the time. I managed to get a supplemental quota for two more people—one for me and another for a friend of mine, Captain Niehoff, who was a Princeton graduate. I chose the University of Aberdeen and Niehoff picked Glasgow. The fall term ran from October 5 to December 20, 1945. Niehoff would be in charge of about fifty soldiers enrolled in Glasgow, while I oversaw the same number at the University of Aberdeen.

In the meantime, our regular duties continued, with some pleasant variations from an otherwise boring routine. I had two girlfriends, Lulu from Lyons and Giselle from Salon—attractive women who furnished some female company in rural Salon.

A trip to Lourdes was another interesting diversion. Our chaplain, a Presbyterian, had a great desire to visit the famous, miraculous shrine located there. With two vehicles—a truck carrying two soldiers, and the chaplain's jeep with his driver and me—we drove across the southern coast of France to reach Lourdes, in the Pyrenees near the border with Spain.

The shrine at Lourdes is an oasis of faith for many who face incurable health problems. The place's fame, dating from 1858, stems from repeated apparitions of the Virgin Mary to a peasant girl named Bernadette Soubirous at a spring in a grotto. Thousands of hopeful pilgrims stream into Lourdes annually, drawn by tales (some of them authenticated) of recovery from medically incurable diseases for those who immerse in the spring-watered pool.

Crutches, braces, and other devices filled the hospital/dormitory near the grotto. These, along with recordings filed in folios, offered evidence of cures that could only be explained as miracles. The chaplain filled the truck with jerry cans of water from the spring at Lourdes to take with him on his long-awaited return to his church in the United States. I, too, was awed by the outpouring of faith and the evidence of its positive manifestations.

As my last official duty as aide to General Rollins, I acted as his tour guide in Paris. Since I had been there before for short intervals, I generally knew what would interest him and the other generals he had invited as his guests. We billeted in one of the finest hotels in the city and had access to its dining room and wine cellar. We toured the Louvre, the Eiffel Tower, Les Invalides, Montmartre, and finally attended the Folies Bergère, which, at the general's request, we visited a second time.

1946

POSTWAR DUTY: ABERDEEN AND AUSTRIA

On October 5, 1945, I received orders transferring me from the Sixty-sixth Division to Vienna, Austria, with a two-month delay en route to Aberdeen, Scotland, while I attended the University of Aberdeen. My official assignment read: Training with Civilian Agency, Information and Education.

The University of Aberdeen, the oldest university in Scotland, was founded in 1495. It favored Scotland with many of its graduates. Its faculty and student body had suffered from five years of war and were somewhat frayed. The arrival of the American GIs gave the institution a welcome jolt. The civilian-soldiers were billeted in family homes and, except for the university noon meal, were pretty much on their own, re-learning the unregimented way of civilian life.

I lived with the Davidson family: George, his wife, and his daughter. Their modest home lacked central heating and the only source of hot water was the kitchen stove. Meals were simple, with lots of tea. "Tea" also referred to the evening meal at around five o'clock. The nights were chilly, but a heated pottery "pig" filled with boiling water made the initial entry into bed less tortuous.

The Davidsons were typical Scots: quiet, industrious, uncomplaining, and friendly in a reserved way. They were practicing Presbyterians but were tolerant of other sects. Aberdeen served as a market for regional farmers, and the main industries were granite mining and milling, fishing, and handling cargo at the port facilities.

My academic career was secondary to my interest in exploring Scotland, England, and Ireland. Every weekend, my friend Niehoff and I ventured to a different locale. We toured the Isle of Skye and had tea with Flora McLeod, the laird of the Clan MacLeod at Dunvegan Castle. We hiked in the Grampian Mountains and climbed Ben Nevis, the highest peak (4400 feet) in Scotland. We toured Edinburgh and Glasgow and played golf at famous St. Andrews. We spent one weekend at Fort Augustus Abbey, Portsmouth Priory's mother house on Loch Ness, experiencing the life of a Benedictine monastery and searching for the famous Loch Ness Monster.

Despite the Scottish dislike for the English, we collaborated with the "auld enemy," spending a long weekend in London. We saw Lawrence Olivier playing in a double header, visited Parliament in session, and had a private tour of the Tower of London, led by a Beefeater who had been a guard there for twenty years.

I also fit in a trip to Dublin. Eire, south Ireland, was off limits to American troops, as it was supposedly neutral during World War II. Despite this, I managed—I can't recall just how—to get orders permitting me to cross the border from North Ireland into Eire. Although I was in uniform, when the train stopped at the border I was inspected and permitted to continue on to Dublin.

Eire was a land of plenty compared to frugal, war-worn Scotland. I stayed at the Salthill Hotel, located just outside of Dublin. A quiet, elegant place, the Salthill served meals varied and delicious. There I met a newly married couple, an Irish bride and a Yankee groom who had served in the Royal Air Force. The newlyweds invited me to their dinner party at the hotel. For the next three days I was toured and feted by several of the dinner guests. Though they were thoroughly Irish patriots, these new friends were very proud of the thousands of Irishmen who had served in the British army. I returned to Scotland with some regrets, wishing I could linger with this cordial and attractive group of Irish people.

By December 1945 my stay in Aberdeen was winding down. By now, the GIs had integrated themselves comfortably in the life of the city and the university and had made many Scottish friends and, in some cases, sweethearts. On December 20, at a farewell dinner and commemoration, the president of the university officially admitted us as alumni of the University of Aberdeen.

Departing from Aberdeen on December 22 was bittersweet. We were leaving friends and academia to return to military regimentation. On the other hand, our departure brought us a step closer to our return to the United

States and civilian life. We waved goodbye to a capacity crowd as our train left the station to head south to London and, ultimately, Paris, the dispersal point where individual soldiers would return to their various army units.

Once in Paris, I was to continue on to Vienna, Austria. Failing to scrounge a ride on an airplane, I had to lay over in Paris and wait for a train. It was Christmas Eve, and I attended midnight Mass at Notre Dame. It was an unforgettable experience. The exterior of the cathedral was aglow, with snow flakes cutting like blinking spotlights through light beams. The crowded interior reverberated with organ music and the singing of the exhilarated choir. Glittering candle flames and the scent of incense created an inebriating sense of good will and warmth throughout the church.

The bright glow of Christmas flickered out on December 27, when I boarded an army train at the Gare de l'Est, bound for Vienna. Twenty-seven hours later, the train chugged into the war-torn city of Munich, Germany, a victim of heavy wartime aerial bombing. Changing trains, the journey continued for twelve more hours, finally reaching Vienna at four o'clock in the morning on December 29, 1945. There, another assignment and a different adventure awaited me.

Defeating the German army and occupying Austria had been the ultimate objective of the Allied advance northward through Italy. Russia, with the same objective, advanced from the east, invading eastern Austria and actually occupying eastern Vienna. Once hostilities ceased, Vienna became the headquarters for the military occupation of Austria. The Fifth Army, commanded by General Clark, became the U.S. occupational administrative unit for western Austria. The British and the French also occupied parts of the country. Russia controlled eastern Austria and eastern Vienna, the area that was directly in the line of march of the Russian army and had suffered the most wartime damage. The United States and Great Britain also occupied zones in Vienna. The headquarters for the U.S. forces in Austria was in an office building in the center of the city.

I reported to headquarters and was assigned to the G-1 Section, which had responsibility for manpower needs for the army and for civilian specialists charged with rebuilding infrastructure in Austria and bordering countries. My boss, a regular army colonel, ex-cavalry, was an old friend of General Clark. The colonel delegated responsibilities and relied on his staff to run the section. Vienna was a coveted post he didn't want to lose, so he demanded top efficiency from his people.

Personnel administration was a new job for me. My assignment included interviewing candidates for replacement officers, who often were sons of senior army commanders. The colonel understood that carefully directed favoritism was a two-way street. I did my job well and in time I became a right-hand man for the colonel in areas other than strictly personnel matters.

As a staff officer, I was not involved in caring for soldiers. I was billeted in a one-room apartment with a sitz bath on Alza Strasse, not far from headquarters. I took my meals in an officers' mess and had no real expenses, and as a very senior first lieutenant, my pay and allowances were more than adequate for a comfortable lifestyle in poverty-stricken Vienna. When not on duty, I was a free man.

The civilian population of Vienna survived on starvation rations. The large Russian army, located in Vienna and across the Danube in Hungary, commandeered all local food production for its own needs. The Viennese subsisted primarily on split peas shipped in jute sacks from Arkansas and rationed out to Austrian food depots by U.S. quartermaster troops. Split pea soup became the staple. Money was practically worthless, since there was so little to buy. Cigarettes became a currency of sorts. Tobacco was a luxury and those who had cigarettes could trade in the black market for special items. In the topsy-turvy economic order, Austrian waiters in the U.S. army recreational clubs enjoyed the status of the nouveau riche, since they had access to cigarette butts, whose bits of tobacco were the equivalent of gold. They rolled the tobacco into new cigarettes that could "buy" what otherwise was unavailable.

Civilian activity centered on finding food and keeping warm, as the city's infrastructure was badly damaged from bombing and artillery fire. Water and electrical systems, traffic lights, elevators, buildings—everything was in disrepair and functioned only sporadically. Civilian employment depended on the military and the whims of the Vienna city bureaucracy.

Life did go on, however. Music was the lifeblood of the city. Ballet, light opera, and concerts served as pleasurable distractions from the harsh reality of everyday living. Audiences, bundled up in overcoats, caps, and shawls, were spellbound by the musicians and ballerinas performing in the frigid theaters. With the advent of spring, outdoor concerts and park activities helped to divert and distract the population from the grim experience of living under foreign military rule.

The allied administration of Austria and Vienna was organized around the four occupying armies: Soviet Russia, the United States, Great Britain and France. The Quadripartite Council, made up of representatives of the four nations, stood at the top rung of the administrative ladder. Chairmanship of the council rotated among the generals representing these countries. The council's decisions and regulations were disseminated and put into practice by the four military bureaucracies concerned.

General Mark Clark served as the U.S. commander. I had known his French-speaking aide, Captain Brie Taylor, in college. When it was the French general's turn to chair the council, French naturally became the language in use, and Brie interpreted for General Clark. Brie became friendly with many of the key players on the council staff, and I benefited from his relationship with them. It was customary for each new chairman to orchestrate an inaugural social affair to celebrate the chairmanship change and, at the same time, to heighten his public relations image. Through Taylor, I attended some of these events.

When Marshal Ivan Konev, the Russian commander, assumed the chairmanship, his inaugural party was unforgettable. I was fortunate to attend, representing my boss, the colonel, who was otherwise occupied. The nighttime extravaganza took place in the throne room of Hofburg palace. This former residence of the Austrian emperor, located in the Ringstrasse in the heart of imperial Vienna, was approached by a long, tree-lined carriageway. A heavy snowfall fell through the windless air on the night of the Russian general's party. A handpicked honor guard made up of tall Mongolian infantrymen stood on both sides of the main entrance of the palace. The guardsmen, spaced at intervals of about ten feet, wore bearskin bushbies and shouldered polished rifles tipped with bayonets gleaming from the spotlights illuminating the way.

A wide marble stairway, lit by crystal chandeliers suspended from the high vaulted ceiling, led to the throne room. Russian soldiers escorted the guests, mostly military men in full dress uniform, to their seats. At the appropriate time, a blaring of bugles announced the arrival of Marshal Koniev and his immediate staff, who paraded to their seats. It was an impressive entrance of a Russian military hero. The marshal had commanded the victorious Russian army in the western offensive that resulted in the defeat of the German/Austrian forces and the capture of Vienna. The climax of this celebration of Russian might was a thrilling performance by costumed regional folk dancers from each of the states making up the Soviet Republic. It was an unforgettable portrayal of Russian culture and political power.

Witnessing these Quadripartite celebrations was an exception rather than the rule, however; it was not part of my normal duties. As Assistant to Colonel Kogstad, the chief of G-1 Officer Personnel, I was involved in estimating and organizing the replacement needs of the U.S. forces in Austria. My main duty consisted of interviewing newly assigned officers of all ranks, determining the most suitable positions for them, and placing them in those positions. In this capacity, I became familiar with most of the army units in Austria and was aware of the pluses and minuses regarding duty in each one. Needless to say, pressure for favorable assignments came from many sources. In the course of my duty, I met and became friends with Captain Clark, Captain Patton and Lieutenant Eisenhower—all sons of the leading players in the U.S. Army.

When I wasn't on duty, I enjoyed hiking in the woods, sightseeing in the city and trying to learn to waltz with my girlfriend, Herta Laimer. I also received three interesting, official assignments outside my normal duties. The first was a three-day stay at the Krupp hunting lodge, which served as the Forty-Second Division artillery officer's RC (rest center). The lodge was located at Blumbach in the mountains about forty-three miles from Salzburg. (Adolph Hitler had sought refuge at his aerie called Eagle's Nest at Berchtesgaden, near Blumbach.) The illustrious Krupp family was one of the wealthiest and politically prominent families in Germany. The hub of the family's steel, iron and weapons manufacturing empire lay at Essen, Germany. At the time of my stay at Blumbach, Krupp von Bohlen and his wife, Bertha Krupp, were living in a guesthouse on the estate. They had left Essen during the allied bombing and after the war were being held as "captives" at the estate. "Big Bertha," the cannon that shelled Paris in World War I, was named in honor of our "hostess," Bertha. (Krupp was later convicted of committing crimes against humanity and sentenced to twelve years in prison.)

The Krupp estate, set amid magnificent mountain scenery, was the epitome of a German hunting lodge. The schloss's walls were mounted with trophies and firearms used in hunts on the estate. The huntsmen employed by the estate were classic, traditional mountain guides. The cross-country skiing was superb. During my stay at the lodge, Krupp's chef served the guests nobly, making my return to army mess a disciplined chore.

Another assignment sent me in March to Rome, the Eternal City, where I spent all of twenty-four hours. Colonel Kogstad had made arrangements for me to fly to Italy on an assignment of some importance to the army and to him personally. Naturally, I was excited and relished the role of a "special agent." My orders were to deliver a packet, contents undisclosed, to General Jaynes, deputy commander of the Rome Area Command. Furthermore, I was to obtain three Medals of Freedom for delivery to the Vienna command and to pick up a set of stamps of the Council of Trent series at the Vatican post office. This last assignment had the highest priority, since the colonel wanted the stamps as a gift for his young daughter, who was an avid stamp collector in the United States.

Armed with official orders, I was driven to the Air Force landing strip outside of Vienna, where I boarded a waiting DC3. The two pilots and I were the only people on the plane. I sat in the cavernous cargo bay. We took off and headed for Compiano Airfield in Rome, but had to take a detour over Trieste, Italy, adjacent to the Yugoslavian border, because of poor flying conditions (swirling clouds, mist, and poor visibility). Alone in the cargo area during the bumpy flight, I was bouncing along and looking out at the stormy weather when suddenly I saw a fighter plane immediately off our right wing. The plane's markings made it clear it was not an American aircraft. It flew in a peculiar pattern, making a sideways maneuver. No sooner had I noticed the plane than we made an abrupt change of direction and headed on a different course. The alien plane veered off and disappeared into the mist.

Upon landing, the pilot cleared up the mystery for me: the phantom aircraft was Yugoslavian; by its maneuvers it had warned our pilot we had strayed off course and were flying over Croatian-Yugoslavian territory. Fortunately, the interception was friendly, as we could have been forced to land in an area controlled by suspicious Russians. I felt touched again by the good fortune that so has often blessed me during my life.

When the plane touched down in sunny Italy, I was driven to my billet, the Hotel St. George, mightily relieved to be in an allied country populated by friendly Italians, rather than surly, hungry Austrians. That night I decided to investigate the Hotel Imperial, which had been the plush rest center for officers on leave from the Italian front. I had heard about it from friends in the Tenth Mountain Division, who had stayed there. Now it was the upscale club for U.S. officers serving in post-war Rome.

I proceeded to the hotel and went directly to the bar, which was stocked with well displayed bottles. There was no Scotch whisky in Vienna, and I hoped the upscale Hotel Imperial could offer me some. In my Italian-accented English,

I asked the suave Italian bar tender if he had any. He nodded vigorously, reached for a bottle, and proudly and professionally presented it, label-up, for my examination. My eyes went wide. Centered in the bottle's green label was a picture of a yellow ball with an elephant balanced on top of it, his trunk raised as if in victory. I read the label: "English/Scotch Whisky, Distilled In Italy—Guaranteed Nutritious And Wholesome." I thanked the bartender and ordered a brandy.

Finding the Vatican post office was something of a challenge. It was not obvious like St. Peter's, but upon inquiring I found the little office and obtained the desired stamps from the ordained postmaster. I could easily have extended my stay in Rome, since there was no priority requiring my immediate return to Vienna. However, I had accomplished my assignments, so I returned with the stamps without any delay.

My third assignment of interest came in May, when I was given a week's leave and joined some fellow officers on a visit to Switzerland. U.S. soldiers on leave there could choose between several activities. My companions and I selected an option that included some hiking in the French-speaking area around Montreux. We stayed at the Montreux Palace, a charming old hotel situated within viewing distance of the Jungfrau, Matterhorn, and peaks of the Dents du Midi. We thoroughly enjoyed the hiking and the luxury of the palace after being stuck in bleak Vienna. A jazz band made up of American Negro musicians provided us with entertainment. The group had been billed at the hotel when the European war started and remained exiled there as a house band during the whole of World War II.

We blew our allowance of fifty Swiss francs after seven days and returned to Vienna, envious of the musicians exiled in the Montreux Palace Hotel.

On June 20, 1946, I finally received orders redeploying me back to the United States for discharge. Obviously, I was delighted, although my excitement was muted by a strange feeling of ambivalence. I had become familiar with the military life and had an uneasy feeling about what might be in store in my life as a civilian. I was awarded the Army Commendation Ribbon, which was gratifying, and I took a train to Le Havre, France, the port of debarkation.

I was returning to the states almost a year after the great redeployment sent most of the U.S. Army home from Europe. Thus, the immense debarkation facilities were almost all vacant and had been poorly maintained. I was assigned to a ship and advised we would depart in three days. I was then assigned to an empty Butler building (a steel, warehouse-like structure) that would serve as my barracks while I waited to leave the port. I was the only occupant of this metal shack. My "fart sack" and pack were the only items on the building's dusty and worn wooden floor. Unrolling my bed roll and hanging my pack on one of the steel support braces, I left the forlorn building to see Le Havre.

Returning at dusk, I entered the building and proceeded to slip into the stretched out sack for a night of sleep. As I inched into the bag, my bare feet encountered a furry blob inside. The creature immediately reacted, retreating to the end of the bed roll. Horrified, I leapt out. I throttled the mouth of the sack shut and frantically beat it against the adjacent steel bulkhead. Exhausted, I shook the sack and out dropped a large, battered, dead rat. Needless to say, I was delighted when I boarded the transport on June 7, destined for the port of New York.

The voyage from Europe back to the United States was a turnabout from the passage to Europe two years previously. Instead of an officer's stateroom, we officers were billeted in the troop area, while the officer's staterooms were filled with European GI-wives or brides-to-be, a varied assortment of women of French, German, Belgian, Hungarian, and other nationalities. Some were anxious, others delighted, some attractive and some otherwise—but in any case the GI bride shipment was strictly off limits.

On July 15 the ship arrived in New York and, amidst all the happy excitement, some of these women found there was no one on shore to meet them. They stood dockside, lonely, abandoned women in a strange land.

In contrast, my college roommate, Joe Downer, was waiting for me. We proceeded to the Harvard Club for a night on the town, celebrating my return after two years overseas.

I returned to Santa Fe on July 28, 1946. My siblings and I all descended upon my parents in Santa Fe that summer. My sister Caroline, a Red Cross volunteer, came back from Japan in May 1946. My brother Hank had already arrived in New Mexico after service as a vice-consul in South America for two years. The "second generation" of kids in our family, Mark and Booker, were home for the summer from boarding school. So, my long-suffering parents were the victims of a complete family return. It was a sustained invasion rather than a brief reunion.

Mother, along with her faithful staff, Josie and Chonita, was summering at the Daniels ranch on the Rio de la Casa, a beautiful and remote sanctuary my father had leased, some twenty miles into the mountains from Mora, New Mexico. All the kids joined Mother at the ranch. This left my father alone like a widower for the summer, batching it at the family home in Santa Fe.

We dubbed the Kelly summer reunion in New Mexico the Return of the Natives. The children and guests disrupted the serenity of this mountain retreat. Our time there gave me a wonderful break from the crowded, regimented military existence I had been leading.

In August I returned to Santa Fe to help my father with several business situations that had developed. I was also eager to put in some time with the company because I had been accepted at the Harvard Business School for the term commencing in February and I needed to earn some money and gain some business experience beforehand.

The first task demanding my attention was liquidating the assets of Jackson Cattle Co., Gross Kelly's livestock operation. Negotiations for the sale of the cattle ranch had been finalized on August 7, with the new owner's occupancy to begin in October. The liquidation would have been fairly straightforward had it not been for the untimely death of the cattle company's livestock manager, Con W. Jackson.

Con's boots would be hard to fill. He was a uniquely colorful character in a profession that featured characters. He had been a cowboy and a rancher for some forty years. There was little he did not know about livestock breeding, marketing, and ranch management, but alas, he was gone.

My father assigned to my brother and me the task of completing the terms of the sale, which involved only the ranch real estate; we had to dispose of the livestock and ranch equipment separately. For two months we labored at the task, painting ourselves into a corner—the corner being the last railroad carload of livestock to be sold.

Con had negotiated the sale of the cattle before his death, leaving us to round up and ship the various lots—bulls, cows, steers, calves—to the buyers. This involved a considerable number of hours on horseback, herding the animals to the railroad shipping point at Chapelle, some twenty miles from the ranch. The ranch cowboys, Miguel and Candido, along with Hank and me, did most of the work. It was hard work that only seemed glamorous after the job was done and a cold beer was in hand.

One inventory item we found hard to sell was a lot of some sixty head of mules that had been bred and raised on the ranch. Con had witnessed the strong demand for military mules during World War I and anticipating a similar demand in World War II he introduced jackasses to the mare horse herd at the ranch. The highly mechanized military in World War II didn't need mules, and by the end of the war, none had been sold. We had black mules, red mules, gray mules, and blond mules. They were all beauties, but no one wanted them.

Fortunately for us, there was a war going on in Greece, with the government fighting Macedonian guerrillas who were ensconced in rugged terrain. The Greek forces needed to transport men and equipment to the combat zone. The U.S. government was supporting Greeks and purchased our mules for them. It was a picturesque scene as this large string of mules wound its way over the range to the rail yards at Chapelle. Little did they know they were headed for the hills of Greece.

The last item we sold were the faithful cow ponies that had served us so well. We sold twenty-five of these trusty saddle horses. We herded our mounts to the shipping point, unsaddled them, and led them into the railroad car. We were now afoot, with horseless saddles, and the job was done.

In mid-August, my father and mother and I attended the Gallup Inter-Tribal Indian Ceremonial in Gallup, New Mexico, the "Indian Capital of the World." One branch of the Gross Kelly enterprise was located in Gallup. Each year at the Ceremonial, Indians from many tribes and reservations around the country gathered there for games, ceremonies, rodeos and other activities. Tribal leaders and Indian traders from the Navajo reservation and elsewhere attended, as well as Boston wool buyers and other commodity dealers. It was important for someone from Gross Kelly to be there to see and be seen. We were hosts at some activities.

On November 28, 1946, my father and my brother Hank attended the Zuni Shalako ceremony at Zuni pueblo. The Shalakos, representatives of the gods, are specially designated Zunis disguised in twelve-foot-tall, bird-like costumes. They descend into the pueblo plaza to dance, race, and bless the pueblo people. Attended by a retinue of mud-heads and other ornately costumed attendants, the Shalakos visit selected houses. The house blessing and dancing last all night. The next day, the Shalakos race in the grand finale of this spectacular ceremony.

The successful completion of a Shalako ceremony portends a bountiful year to come. In contrast, should some mishap occur during the ceremony—one of the racing Shalakos suffering a fall or injury, for example—it is seen as an omen of ill luck for the spectators and the Pueblo.

That year my father and Hank witnessed a race where one of the Shalakos fell. The crowd immediately melted away and gloom settled on Zuni out of a dread that ill fortune would haunt the Pueblo and its people until the next Shalako appearance.

Caroline left New Mexico for Washington, D.C., to line up a job. While there, she suffered a nervous breakdown that prompted my father to go to Washington. The doctors opined that Caroline should be hospitalized and she was admitted to the Menninger Clinic in Kansas for psychiatric treatment. My father's comment was, "The Zuni portent has certainly hit us hard."

Family, cousins, and friends attended Christmas dinner at my grandmother's house back in Santa Fe. As 1946 came to a close, I prepared to start my studies at the Harvard Business School.

1947-1950

HARVARD REVISTED AND A RETURN TO NEW MEXICO

Harvard Business School

I had applied for admission to the Harvard Law School while I was in the Army, but was turned down. At that point, I directed my efforts toward entry into the Harvard Business School. I was admitted and commenced the program in February 1947.

I returned to Harvard after a three-and-a-half year hiatus. The business school, like the law and medical schools and other departments, made up Harvard University. Harvard College was the undergraduate seed bed for the university's several graduate schools. The business school students, in contrast to the college undergraduates, represented a sweep of Americana. Most had served as commissioned officers in the army. They were ambitious and highly motivated, intending to gain maximum advantage from attending the renowned Harvard Business School—and then to hit corporate America on the run.

I somehow felt I was owed the admission to the school. I had a more laid-back attitude and to some extent I felt my new classmates should automatically recognize my Harvard College superiority and treat me with deference. I soon realized my sense of superiority was misplaced and that to survive in this arena academically, I would have to "root hog, or die."

I was assigned to Mellon Hall, named for a financial titan who had been a major contributor to the school. My roommates included Anders Sundberg from Stockholm, Sweden; Hugh Richardson, a regular army major from South Carolina; and Peter Houwsam from Ohio, who had the warm and persuasive characteristics of a born salesman. Amid this diverse, high-powered mix, I wondered why I wasn't still in New Mexico. In time, my roommates and I bonded and became a friendly and cooperative team.

The business school taught by way of the "case method," studying actual situations and problems that had occurred in various corporate organizations. We considered problems associated with the spectrum of business activities, such as accounting, finance, personnel, marketing, manufacturing, sales, and engineering. We discussed and analyzed these cases and offered solutions. Our examinations involved cases we had not previously studied and our grades were based on our analyses and solutions.

I found the curriculum extremely challenging. The problems we had to solve in army training had emphasized trial and error. For them, there was always a straightforward answer—the approved "army school solution." At the business school, in contrast, there could be several answers for any given problem—or there might not be any solution at all. This frustrated me greatly.

The workload at the business school was heavy and demanded frequent written reports. Accounting, finance, and statistics proved to be tortuous subjects for me. I soon became a problem student, prompting a committee of deans to call me in for an interview. They recognized it wasn't a problem of not striving, but rather one of approach and comprehension. Through their help, I managed to attain a respectable, though not outstanding, record.

Extracurricular activities provided me necessary relief from the demands of the program. Sundberg, the Swede, knew a bevy of Swedish and Norwegian female exchange students around the Boston area. These charming and good-looking women, somewhat freed from restraints of Nordic society, were great pain relievers. Also, squash and running kept me in fair shape. Skiing, another passion of my Norwegian friends, offered diversion on an occasional weekend. We abused one type of recreation—drinking. This habit may have been a holdover from the military, and it was a problem on free weekends, requiring Monday morning hangover medication.

While I was in Boston struggling with hypothetical business problems, the omen of the fallen Shalako kept hammering adversely at my family in New Mexico, disrupting my studies.

My brother Hank, an adventurer and gifted writer, had been planning a journalistic float trip down the Rio Grande from its source to the Gulf of

Mexico. On Mother's Day, May 11, 1947, while in training for the planned adventure, he and his partner, ex-Captain Murphy, U.S. Air Force, were tossed out of their rubber raft and drowned in the swollen, surly spring runoff of the Rio Grande in the gorge near Pilar. It was a major tragedy.

I flew out immediately when my father called with the news. I joined the search party scanning the turbulent brown river the day after the accident. Two days later, Hank's battered body turned up on a sandbar some eighteen miles from the accident. Murphy's body was not recovered until several weeks later, when it was found near the accident scene entangled in some riverside brush. Following Hank's funeral, I returned to Boston and worked extra hard to catch up on the missed subject matter. The Shalako bad luck had struck again.

In contrast to that dark tragedy, a happy event took place on September 2. My sister Yei married Frank D. Gorham. I flew out to Santa Fe for the wedding, which coincided with the annual Santa Fe Fiesta. Frank and my college roommate, Joe Downer, had served as pack artillery officers in the Tenth Mountain Division, a unique unit in the army. The division gained fame in Italy, and Frank was awarded the Silver Star for heroism in combat. After the war, he completed his studies as a petroleum geologist, all the while courting Yei from a distance. At the time of their marriage, he was employed in Venezuela as a geologist for Gulf Oil. (He and Yei lived in Venezuela for five years before returning to New Mexico.) After the wedding and fiesta, I returned to the land of debits and credits, recharged for the continuing academic battle.

In November my father came to Boston on a business trip. He met and, as expected, wined and dined my roommates. He also joined me and two of my college classmates, Downer and Sharpe, and their fathers for the annual Harvard-Yale game.

During this pleasant visit, my sister Caroline, a patient at Menninger Clinic, suffered a relapse. This triggered a period of soul searching for my father, who decided to transfer Caroline from Menninger to a psychiatric hospital in St. Louis, where her uncle and aunt, Pip and Pat Gross, could monitor her condition.

Another event in New Mexico in the middle of December caused me to return to New Mexico again. Jay Van Soelen, the beautiful young daughter of our friends the Van Soelens in Santa Fe, suffered fatal injuries in an automobile accident in Albuquerque. She lingered unconscious for almost ten days.

Her fluctuating condition tore at the heartstrings of her parents and their friends. My parents booked in at the Alvarado Hotel in Albuquerque to support the Van Soelens during their trauma. They asked me to come home to join the watch. I was a pallbearer at her funeral on January 22. We hoped this was the last move of the Shalako portent.

The fall term at the business school ended while, at Zuni, a safe and successful Shalako race at the end of the year gave us an optimistic outlook for the coming year, 1948.

Now an experienced graduate student, I faced six more months in the program before I would venture out into the world of higher commerce. The route to the business world was well established: more classes, research, visiting different businesses to observe their operations, and interviews with possible employers.

During a week of vacation in February, I visited the Downer family on Long Island and called on several banks and other businesses in New York City. In May I was in Baltimore to investigate an employment opportunity. I had become intrigued with the B. F. Shriver Co., a farming concern located in the beautiful country of western Maryland. I was related to the Shriver family through my father. The family owned and managed about six thousand acres of farmland, producing vegetables such as canning peas and carrots and other food products. In addition, B.F. Shriver raised cattle for the premier meat markets in Baltimore.

The Shriver lifestyle appealed to me. These gentlemen farmers and food manufacturers had a long historical connection with agrarian social circles in Maryland. The mix of rural life, farming, and the horsey social life, plus a profitable business opportunity, made a management job there attractive to me. The company offered me a position as assistant to the president, which I took under advisement, pending further interviewing opportunities.

A job opportunity with Eastern Air Lines also was quite interesting and flattering. I cannot recall all the preliminaries that led up to the job offer, but the climax was a private, hour-long visit with Captain Eddie Rickenbacker, the president of the airline. At the time Eastern was the premier U.S. airline. It maintained profitable runs from New York City to Florida, with additional routes branching off this golden flight line. Next to Pan American, renowned for its extensive coverage in South America, Eastern was the jewel of the airways.

Captain Eddie had earned fame in World War I as the leading fighter pilot/ace of the Army Air Corps. He had gained his experience and reputation as the leader of the Lafayette Escadrille, a U.S.-manned French fighter squadron flying the Spad fighter. After the war, Rickenbacker started Eastern and rose to the top of his profession. My interview was focused on my qualifications for and interest in a position with Eastern, but as I recall, our conversation wandered to his war experiences as he showed me his World War I flying gear and his many decorations.

Standard Oil of New Jersey and the Chase Bank also interviewed me, but I cannot recall the outcomes. I think they offered me jobs, pending the successful completion of my program at the business school. I do remember juggling several job possibilities and at the same time feeling a magnet pulling me westward, towards New Mexico.

I took my final exams May 19 and graduated June 10. Now that I had successfully passed I could show my enviable Harvard MBA on my résumé. It was a commendable achievement, but I could not decide what employment path I wanted to take. My indecision was frustrating and complicated by the lure of employment in our family company. Working for Gross Kelly & Co. would be familiar and it would allow me to live in my favorite part of the world, not to mention the advantages of being a member of the owning family. However, I was well aware that Gross Kelly was an old company, riddled with nepotism and some long-time employees who were performing marginally. The stockholders, mostly non-resident widows and children of former owners or managers, depended for their income on the dividends, which became harder and harder to earn. Competition was closing in and the capital needed for modernization was difficult to obtain. The challenge of correcting all this would be awesome, but the job was there for the taking.

Returning to New Mexico and Working with Gross Kelly & Co.

My interest in the Gross Kelly business lay in livestock production, timber, and wool processing and marketing—areas once profitable for the company but now in decline. Now the business focused on wholesale grocery and hardware sales and distribution. My indecision led to a period of dabbling, searching, and toying with the possibility of getting involved with farming, ranching, and other outdoor pursuits. I kept close to the company but avoided being consumed by it. Perhaps part of my quandary came from the fact that I was still adjusting to civilian life after a considerable time in the military.

In any case, I relocated to New Mexico, but as far as my career was concerned, I was like a hummingbird flying from honey pot to honey pot in search of the perfect nectar.

On reviewing my 1949 diary, I am amazed at the variety of activities that engaged me, as well as my perpetual motion and the freedom I was given, or assumed, during the year. In the end, I decided to follow the suggestion given me by George Bloom, then the president of the First National Bank of Santa Fe. George was a close family friend and a wise and patient practitioner of listening and then offering simple yet practical advice.

George was well aware of the problems facing Gross Kelly & Co., as he was the company's primary local banker. I had visited with him frequently, discussing my reservations about joining the Gross Kelly management. He suggested I give the company a nine-month trial, working at all the branches and in every department, and then leave it and work in an allied business for a year. My tolerant father approved of George's recommendation. He understood my hesitations but, deep down, he wanted me to work for the company. I think he hoped an employee with an MBA from Harvard could work a corporate miracle.

On March 30, 1949, I was officially employed in Gross Kelly & Co.'s general office. There was no job description; I was simply to learn the business. During the following year, I spent several months in the lumber division in Las Vegas, learning how timber was extracted from the forest and how the logs were graded for use and milling; I also studied the basic procedures required to create finished products. I liked the work, but I realized early on the operation needed to be mechanized to be more cost efficient.

From Las Vegas, I transferred across the state to the Gallup branch of Gross Kelly. This operation was a cross between a modern wholesale grocery distribution operation and a bartering, commodity-oriented trading operation with the Navajo, Hopi, and Pueblo Indians of western New Mexico and eastern Arizona. The branch's market area ran from Grants, New Mexico, to St. Johns, Arizona, and from Gallup north to Farmington and Durango, including the vast and sparsely settled Navajo Reservation.

In order to properly supply the modern supermarkets of Gallup and Farmington and, at the same time, the isolated trading posts of the reservation, a diverse inventory was required. We needed to supply supermarkets with the usual case lots of grocery store foods items. In contrast, the trading posts required a most diverse range of inventory. This included dry goods, shoes, blankets, clothing, camping items, livestock feed, rope, saddles, tanned leather, patent medicine, paper goods, and soda pop, as well as the usual staple grocery items.

To further complicate the economics, the supermarkets required credit, billing, and normal cash transactional accounting, while the traders on the reservation offered commodities they acquired in trade from the reservation—sheep, wool, hides, piñons, woven blankets and jewelry—in lieu of cash, pledged as collateral for the supplies obtained from the wholesale distributors who serviced the trading posts. Gross Kelly wound up with railroad carloads of piñons, wool, sheepskins, cowhides, and large quantities of Indian blankets and jewelry. This in turn demanded that managers at Gross Kelly have a thorough knowledge of the broad commodity market to ultimately turn these products into cash.

The Harvard Business School did not offer courses in commodity trading, Indian ceremonies or how to grade raw wool, cowhides, or Indian jewelry. My time working with Gross Kelly in the Gallup area was a unique experience, to say the least.

In addition to these and other branch operations, I became involved in labor negotiations, financial matters and the necessary civic pro bono requirements of an established business. Generously sandwiched between these business activities was a wide variety of non-business pursuits. I was an active skier and I hunted and played golf, but I met my most serious distraction in Jeanne Wise, who years later would become my wife. She came to Santa Fe from New York for a two-week visit, which she extended to a six-month stay. Our time together that year was the beginning of a lifetime relationship.

I stayed active all year but still questioned the sanity of becoming entangled with the family business. Even as I learned about Gross Kelly's inner workings, I considered many different lines of work and remained hesitant to make a real commitment to the company. Then came some major interruptions that gave me an excuse to put off my decision for even longer.

Sometime in 1949 I had heard the army was seeking commissioned officers to act as tactical umpires for proposed war games in the Caribbean. The assignment intrigued me. I was in the throes of career indecision and the thought of visiting a tropical island appealed to me. Furthermore, my wartime army friend, Brie Taylor, suggested we both apply for the assignment. We completed the necessary forms and applications in a long paper trail towards possible approval.

I kept working and romancing in New Mexico and did not seriously expect a favorable response to our somewhat tongue-in-cheek applications. When the orders arrived in mid-winter in New Mexico, the two-month tropical adventure was especially alluring. The umpire detachments, some three hundred officers and five hundred enlisted men, assembled at Fort Lee, Virginia, where we trained for a month.

The maneuvers involved an amphibious assault by Navy and Marine forces on Vieques Island, which was to be defended by an army division and the Puerto Rican National Guard. In order to umpire the action properly, the umpires had to be trained in amphibious tactics, beach command and control, and the usual infantry, artillery, and air force activities.

On February 16 we embarked on a four-day voyage from Norfolk, Virginia, to the maneuver area at Vieques Island, about twenty miles south of Puerto Rico. The unoccupied island had long served as a target area for naval gunnery training. A high ridge ran like a backbone along the island above the sandy beaches that extended along the island's length. Except for the ridgeline, the beautiful island was covered by palm trees and tropical brush. Our tents were located on the ridge tops and had the benefit of the cooling trade winds that blew from offshore across the high ground.

The war games lasted about two weeks. The army and the Puerto Rican National Guard successfully defended the island. After "hostilities" ceased, Taylor and I were given leave to visit San Juan, the historic capital of Puerto Rico. We were cordially entertained by federal Judge David Chávez and his attractive daughter, Caroline. Both were natives of Santa Fe. Caroline took us sightseeing in the daytime and showed us the nightlife after dark. I held out the hope that the army would release us from duty in Puerto Rico, but in true army form, we were shipped back to Norfolk. Our release date fell on my birthday, March 22.

The Puerto Rico experience had provided an interesting and exciting interlude from the travails of dealing with the problems endemic in Gross Kelly & Co. Now I had to return to them. My job as the operations manager involved overseeing the modernization of the warehousing and distribution procedures. This demanded considerable travel between our several branches, as well trips to other regions to inspect and observe the way comparable companies were streamlining their operations.

Concurrently, Gross Kelly had decided on a risky expansion program that involved constructing a warehouse in Roswell and conducting a market study of eastern New Mexico. Financing this expansion demanded I firm up relations with the several banks Gross Kelly worked with. This, required travel to St. Louis, Chicago and New York City. Since Jeanne lived in New York City, my trips to the eastern seaboard were motivated by more than business finances.

The pace of business activity at Gross Kelly increased, pulling me more and more into involvement with company operations. All the while, I harbored doubts about my commitment to the business. I had always enjoyed ranching and kept up that interest by maintaining a sideline agricultural venture in the Estancia Valley, the Valley Irrigation Company. On weekends and other times, I worked with the sheep and cattle on the irrigated pastures of this company.

That pleasant summer suddenly ended suddenly on June 25, 1950, when North Korean forces invaded South Korea. On June 27 President Truman ordered U.S. forces to support the South Korean army and the United Nations authorized a police action against North Korea. General Douglas MacArthur was designated the force commander.

On August 18 I was ordered to active duty with the army. I wondered whether my Vieques Island venture might have triggered my early recall into the military. (I eventually realized there was no connection; I simply was a reserve officer and was therefore called to duty early in the conflict.)

Because of my considerable involvement in the modernization of Gross Kelly, an effort critical to the business's survival, the board of directors petitioned for a deferment for six months for me. The Army granted it, pushing back my active duty date to February 20, 1951. For the rest of the year I continued focusing my efforts on modernization of the business, enjoying my weekend ranching and the usual social activities as a bachelor.

On February 22, 1951, I reported for active duty at Ft. Sill, Oklahoma.

1951-1952

KOREAN DUTY

The next eight months unfolded like a Charlie Chaplin nightmare. I was a captain on active duty and, at the same time, the hoped-for savior of our troubled family business. I felt ill equipped to fill either role.

As ordered, I reported to Ft. Sill, the headquarters of the army field artillery, for further assignment. For three weeks, I was given various jobs, such as an investigator for court martial cases, duty officer at the post stockade, and troop commander for trains delivering soldiers to various posts.

Permanent assignment orders finally arrived, posting me to the 598th Field Artillery Battalion at Camp Polk, Louisiana. This was an all-black unit, training with 155 mm Long Tom cannons. The town nearest to Camp Polk was Leesville, a small, redneck agrarian village adjacent to the Big Thicket separating Louisiana from Texas—hardly an ideal location for the several thousand black soldiers assigned to the post, or for the white officers commanding them.

After several days of orientation, I assumed command of the Headquarters Battery, the unit that served the battalion commander and his staff. I was the only officer in the battery and the only non-African-American. To make my position more awkward, I was green and therefore totally dependent on the army-wise noncommissioned officers I was supposed to lead. The initiation was not without pain.

When I took command of the battery, the key leaders were either absent or distracted. The first sergeant was AWOL (absent without leave or permission). The supply sergeant was a patient in the post's psychiatric ward. The mess sergeant was having a domestic tiff with his "significant other," who happened to be the ration breakdown sergeant. My company clerk, a huge man whose finger nails were painted, had as sub-clerks a coterie of his boy friends. These noncommissioned officers were the "leaders," in charge of some two-hundred young, African Americans comprising the battery. The non-commissioned officers were the rotten apples in a bunch of otherwise decent apples. I did my best but was not the commander this unfortunate battery needed. (Postscript: President Truman later eliminated segregation in the military. This did more good for the black soldier than emancipation had done in the previous century.)

To compound my problems at Camp Polk, Robert Gross, Jr., the executive vice president of Gross Kelly & Co., died from a heart attack at the age of forty-two. I had been elected a director of the company prior to my army call and after just one week as the new battery commander, I had to request an emergency four-day leave. It was granted and I attended the funeral. Afterwards, at a director's meeting, I was elected vice president of the company. This made little sense to me.

I returned to Camp Polk and resumed my traumatic job as battery commander. Meanwhile, the directors of Gross Kelly initiated efforts to have me transferred to New Mexico or released from the army. This was embarrassing for me and an inconvenience to the battalion commander, who had enough problems without worrying about some businessmen in New Mexico. I felt like a punching bag being drilled by twin boxers.

Presumably, there was a need for replacement officers in Korea, but I'm sure my colonel also wanted to be rid of his problems with this Captain Kelly, whose schizophrenic, hyphenated identity (soldier-businessman) made his life difficult. In any case, I received overseas orders to report to Ft. Sill to attend the basic officer's refresher course in artillery and then to proceed to Korea.

After the month-long course, I was granted a pre-overseas leave of thirty days. By this time I was thoroughly brainwashed, reconciled to doing combat duty, and looking forward to a pleasant leave.

Fearing the worst (imagining me as a combat casualty), the company directors made a last-ditch attempt to reverse the gears of the military machine. My Uncle Francis "Pip" Gross, an active Gross Kelly director, had served in President Truman's artillery battery in World War I. His sergeant then was Harry Vaughn, who was now a major general and President Truman's personal aide and comforter. Pip felt he could use the "old soldier balm" on General Vaughn to win my release from the army. I was an unwilling party to this scheme, but acquiesced to the effort.

World War I Private Francis Gross and I flew to Washington, where we had a cordial visit with the general, full of reminiscences about World War I. Pip pleaded our case. Vaughn was attentive but remained noncommittal, suggesting we pay a visit to Senator Dennis Chávez, the senior Democratic senator from New Mexico. We made an appointment with Chávez and got in to see him, but there was a problem: my father and grandfather, staunch Republicans, did not approve of the senator. Chávez was cordial but somewhat stiff and formal during our interview. Three weeks later I was in Seattle boarding a plane for Tokyo, en route to Korea.

The orders sending me to Korea freed me from the penal duty with the artillery battalion at Camp Polk. Additionally, the tour gave me some distance from the depressing and severe business problems wracking Gross Kelly & Co.

Upon completion of my pre-overseas leave, I flew to Seattle and was processed for assignment to the Far Eastern Command headquarters in Japan. On November 6, courtesy of Canadian Pacific Airways, the plane took off for Tokyo, loaded with individual replacement officers. It was a comfortable flight. Two attractive airline hostesses attended to our needs during the eighteen-hour flight to Japan.

The plane refueled at Shumia in the Aleutian Islands. Early on November 8 Mount Fuji, the magical Japanese volcano, appeared on the horizon. Around noon we landed in Tokyo and were bussed to Camp Drake for processing and further assignment. There I received my orders assigning me to the 999th Field Artillery Battalion of the Third Infantry Division, I Corps, somewhere in Korea.

In contrast to the plush airline flight to Japan, the transportation to my destination in Korea was primitive. A day's train ride took me to the Japanese port of Sasebo, where, after a two-day layover, I boarded the tramp steamer *Kean Maru*, destination Pusan, Korea. With my infantry "fart sack" as bedding, I spent two days "resting" on the rusted steel deck in the cargo compartment of the derelict craft. On November 14 we docked at Pusan. The worn, gray and abused harbor area was overused and under-maintained. As we disembarked, it was raining and the whole scene was one depressing preview of a country at war.

The next stage of the journey involved spending two days on a four-car train pulled by a coal burning, smoke belching locomotive, heading north to the combat zone. For seats the railroad cars had only wooden benches. There was no heat, no water. Glassless windows looked out onto a damp and foggy countryside.

The train ride ended at Yung Dung Po. It was still raining as six poncho-draped, forlorn looking, black enlisted soldiers and I loaded onto the truck, a wet six-by-six from the 999th Field Artillery Battalion. My morale, which had been marginal up to this point, dropped to zero when I saw these fellow passengers. I realized I was destined to serve in another colored unit, positioned in the Never Die Valley on the Korean front.

My reception at battalion headquarters was as cordial as conditions permitted, but the unit did not need another captain. All those slots were filled, but luck was with me. Headquarters I Corps Artillery was short one position and I was ordered to report there for duty.

The United Nations military headquarters, commanded by General MacArthur, was located in Japan. The army in Korea was led by a U.S. general. In turn, the front was controlled by two army corps containing several divisions, each with its organic division artillery units. In addition, the corps had its own artillery battalions, commanded by a corps artillery commander. These corps units reinforced the organic division artillery and, in addition, concentrated on destroying the enemy's artillery capabilities.

I was assigned to the headquarters of I Corps Artillery, which was at a separate location from I Corps headquarters. I served in several capacities that afforded me considerable opportunity to become familiar with the activity in the I Corps section of the front. I served as an assistant combat intelligence officer, as an assistant operations officer, and the I Corps artillery historian. In addition, I was the briefing officer for the various high brass commanders visiting our section of the front. The Intelligence Section (S-2) kept the command informed on hostile artillery capabilities and activities, while the Operation Section (S-3) ordered the actual target designation and firing of the batteries. These jobs required frequent written evaluation reports, and my Harvard Business School training helped in this regard, as well as in the briefings.

When I reported to I Corps Artillery headquarters, I was assigned as assistant to Major Thompson, in charge of the Combat Intelligence section. My orientation to the new job was brief. The major welcomed me and announced he was departing in two days for the United States for reassignment. He added that the monthly combat intelligence summary report was due in higher headquarters in a week and that I was now the active intelligence officer and therefore responsible for submitting the report. This came as something of a shock to me. My platter was full of undigested intelligence information, but now I found myself in the role of the expert, expected to produce a cogent intelligence statement.

I busied myself gathering data on hostile artillery. This involved obtaining and collating information from all our front line units regarding their exposure to incoming artillery shells. There were three divisions occupying the I Corps front, each subject to hostile shelling. Needless to say, I was dependent on each division furnishing me with good information. Somehow, I compiled the report and submitted it on time.

I had no crystal ball, but my analysis met with approval. Several days later, a memo was received with the comment that my report was "superior." In fact, I was told it was the best report higher command had ever received. I appreciated the compliments, but was also surprised my hastily done assessment received such accolades. I realized intelligence information is fragile and subject to errors at every stage of its development, from data gathering to analysis to reporting. I was all too aware of the potential for error and the possibility that higher command could be exposed to inaccurate and slanted information.

While in Korea, I had one, five-day "rest and rehabilitation" leave in Japan. It was neither restful nor rehabilitating. Tokyo was a typical troop destination, complete with bars, dance halls, and geisha girls. I stayed three days at the Royal Fujian Hotel, a resort not unlike the Hotel Coronado in La Jolla. My female golf caddies were adept at recovering valuable golf balls I directed into numerous water hazards.

During my Korean tour, my younger brother Mark was stationed in a quartermaster depot in Yung Dung Po. He had the high distinction of being a private. We both obtained three-day leaves and I was his guest in Seoul at the enlisted men's leave hotel affectionately known as the Frozen Chosen. ("Chosen" is an outdated, Japanese-derived name for Korea.)

In contrast to the creature comforts of Hotel Frozen Chosen, facilities at my headquarters were primitive. While we enjoyed hot, indoor showers on a daily basis at Hotel Frozen Chosen, back at headquarters we took a Jeep trip once a week on a dusty road to a "quartermaster mobile shower unit." There, we bathed in a communal shower in a large tent with a wood-plank floor. After soaping up and rinsing, we were issued fresh underwear and fatigues, only to climb back in the Jeeps for a dusty ride back to our showerless units.

I recall one time when things went awry on our scheduled shower day. We arrived at the mobile shower unit, situated along a stream. As usual, a group of Korean women was scrubbing down their laundry in the river. The shower unit was filled with naked, soapy soldiers when, suddenly, rogue artillery rounds exploded nearby. Spooked by the shelling and afraid a shell would hit the water tank and cause us to be scalded with hot water, we frantic, naked men ran out of the tent *en masse*, to the great amusement of the Korean women. Needless to say, the quartermaster shower unit was relocated.

My Korean tour finally came to an end in June 1952. By then I had been promoted to major. Upon my departure on June 15, I was awarded an Oak Leaf Cluster to add to my World War II Bronze Star. After all, I apparently did a good job while in I Corps. On July 2 I boarded the transport *USS Brewster* in Sasebo, Japan, for the long voyage home.

Our two-week cruise across the broad Pacific to San Francisco was a classic voyage. We three officers sharing a stateroom had no duties to perform. We spent the days reading, sunbathing, playing chess, watching movies, and shooting the bull, all the while mesmerized by the undulating swells of the Pacific Ocean.

As the voyage neared its end, I began wondering what civilian life would be like. What about marriage? I had been thinking of Jeanne. Her letters had been warm and supportive. Should I slip into the familiar harness of the Gross Kelly business, joining the aging team there. Or would I strike out on my own? Where did I want to live? How would I support myself?

The thrill of crossing under the Golden Gate Bridge and docking along the army pier in San Francisco temporarily banished the heavy thoughts of adjustment and they were replaced with anticipation of the pleasant reunions as we trooped down the gangplank onto U.S. soil.

Following two days of orientation and uniform changes, we received orders directing us to our discharge posts, and we were on our way. On July 17 I arrived in Albuquerque and was greeted by a large contingent of family, relations and friends. Mother and I drove to Fort Carson, Colorado, and on July 23, 1952, I was honorably discharged from the U.S. Army.

The rest of the year can only be described as a lengthy adjustment back to civilian life. On the surface, I was occupied with a training position with Gross Kelly. I attended management meetings, accompanied by my father and other administrative officers, and took trips to the company's various branches. I met our bankers and dealt with labor and other personnel problems. But mostly I did what I wanted to do. I visited the Valley Irrigation Company, played golf and tennis, went hunting, and joined in the active social life of my group of friends. I attended the Gallup Indian Ceremonial and was an active participant in several events and parties connected with the Santa Fe Fiesta.

As the year came to a close, I began to realize the honeymoon was over. I could not play the role of spoiled son of the president any more. I had to bite the bullet and do my best to live up to the expectations and responsibilities I had been raised for and trained to assume. Gross Kelly & Co. was hemorrhaging and in dire need of remedial action, or it would simply collapse. I was the responsible doctor on the case, though not necessarily equipped to effect the needed cure.

Bud and Nate Greer at Field Artillery Officer Candidate School, Ft. Sill, 1943

Bud as Newly Commissioned Second Lieutenant, 1943

Battery A Party at Camp Robinson, Arkansas, 1943

Bud and Pilot, Aerial Observation Crew, France, March 1945

Bud and Lulu from Lyon, 1945

With Russian Soldiers, Vienna Woods, During Occupation, 1946

Bud (center) on Leave in Switzerland, ca. 1945

Bud and Badger at Rio de las Casas, ca. 1946

Old Time Cowboy Ruperto Alarid, 1946

I Corps Artillery Headquarters, Korea, 1951

Eight-Inch Howitzer Gun Crew, Korea, 1951

Captain Kelly with Colonel Jack Kenny, Forward Area, Korea, 1951

1953-1959

MARRIAGE, LAUNCHING A CAREER AND DOMESTICATION

1953: Courtship and Marriage

I think of my parents when I reflect on the flow of life during 1953. Dad was sixty-seven, and Mother, who went by the nickname Ranny, was around sixty. Their home on Palace Avenue was still the magnet for their five children. Hank's widow, Dorothy, and her two infants lived in the guest house. My sister Caroline, swallow-like, migrated back and forth, depending on her job dissatisfaction. Yei and her boys were living in Albuquerque; she often commuted to the parental nest for relief from her hectic family life. Mark was about to return home from Korean army service. Booker was at Harvard but came home in the summer. Now foot-loose and thirty-one, I was living at home when not on some personal adventure or activity connected with Gross Kelly & Co.

My parents were constantly entertaining friends, out of town visitors, business people, and those connected with my father's numerous pro bono involvements in organizations. In addition to serving as president of Gross Kelly & Co., he was the president of the New Mexico Museum Board, a director of the Carrie Tingley Hospital, chairman of the St. Vincent Hospital building effort, and director of the First National Bank and the Public Service Company of New Mexico. Mother was busy with her own activities, not the least of which was running the perpetual motion household at 531 Palace. Despite the pace and attendant trauma, my parents orbited in an interesting and diverse galaxy, often to their own exhaustion.

I was involved either directly or indirectly in many of my father's activities. He was skilled at delegating details, often to me. In a sense, I was on call. My assignments with Gross Kelly varied widely: visiting bankers, checking on business problems, mollifying disturbed customers, negotiating with labor organizers and attending meetings as my father's representative. Gross Kelly was always undercapitalized and thus relied on a covey of banks to permit borrowing from one bank to pay off another. This rotation banking required maintaining very close relationships with banks in Denver, Kansas City, Chicago, and New York. My father usually controlled this aspect of the business, but I had learned the drill from him and made several solo banking trips in 1953.

While mending fences with banks I was able to see Jeanne Wise, who lived in Manhattan. I had met Jeanne in the summer of 1949, when she was visiting Santa Fe with a family friend, Claudia Bisell, and met my sister Caroline, who introduced her to me. I was immediately impressed. Here was an attractive, responsive, smart-looking eastern girl who was fun to be with. She was athletic, a good dancer and was quickly popular with my friends. We became a pretty constant pair.

Jeanne grew up in Scarsdale, New York. Her parents were ambitious, friendly and prosperous New Yorkers. Her father was chairman of the board of McFadden Publications. They had three children. Bob, a Culver graduate and Brown University student, had been an officer in the First Cavalry Division. He died of battle wounds in the Philippines. Jeanne went to Rogers Hall, a boarding school, and graduated from Pine Manor Junior College. From there, she joined the U.S. Navy WAVES and was stationed at the Naval Air Training Base at Pensacola, Florida, where she taught aerial gunnery. Undoubtedly, she was a popular target for the naval flying cadets. At the time I met her, she was a broker-in-training at Eastman Dillon & Co. in New York City. Her younger sister Betty was the private secretary of John Watson, the president of IBM.

Jeanne worked for Jimmy Russell's insurance firm during her extended visit to Santa Fe in 1949. She also joined with the active group of young war veterans, of which I was a part. We played golf, tennis, fished, had picnics and parties, went to Indian dances, and, when not working, were constantly on the go.

Although she would have been willing to wed at that time, I wasn't ready for marriage. I was still too emotionally unsettled to make the commitment.

She gave up trying and returned to New York and her financial career. Later, my sister Caroline was working for *The New Yorker* and shared Jeanne's Madison Avenue apartment. Thus I kept in close contact with Jeanne on my trips east until I was recalled for duty in Korea in 1951.

At a farewell party Jeanne gave for me the night of my last visit before Korea, one of her female guests lured me away. Obviously, this didn't please Jeanne. The next day I was dead meat, hung over, exhausted, and banished by Jeanne. I departed for Korea under an interdict. Fortunately, my several letters from Korea softened her heart and on my return in 1952 our relations were warm again.

I made a banking trip in June 1953 that coincided with my tenth class reunion at Harvard. It was the first gathering of my class since the war and I saw it as a good chance to invite Jeanne to Cambridge for the festivities. We had a great time there, golfing, sailing, watching baseball, and dancing and romancing. Knowing the Santa Fe Fiesta in late August would be a chance to see her again, I invited her to Santa Fe for the event, and she accepted.

With Jeanne as a family guest, we attended the Gallup Indian Ceremonial and then participated in Fiesta, with its usual parties, parades and frivolity. The following weekend, September 5, my cousin Mary Ellis Kane and her husband Joe Stein gave a house party at their ranch on the Rio la Casa near Mora. Jeanne and I were guests. I knew the hour was at hand to put up or shut up. That Sunday Jeanne and I went into Mora for Mass. During the service, I received a spiritual order to act, and after Mass I did: I proposed and Jeanne accepted. Now the fat was really in the fire.

Jeanne knew her man and left me little time for reconsidering. She wanted a short engagement, so we scheduled the wedding for November 28, 1953, in New York City. Jeanne was practical; she had known for some time she wanted marriage, but may have been too long-suffering and sensitive to expose her inner feelings and opinions openly. Once I proposed, though, she wasn't hesitant to express her wishes.

On September 12 the annual meeting of Gross Kelly & Co. took place. My father, long anxious to pass on the baton, resigned as president. He assumed the new position of chairman and I was elected president. For the next two and a half months, I was deeply involved learning the presidential ropes and preparing for the wedding.

On November 20 I was the willing victim in a stag party given by my old friends at El Nido restaurant in Tesuque, and on November 24 I departed for New York for the nuptial festivities. Jeanne's parents pulled out all the stops. The rehearsal dinner took place in the Maisonette Room at the St. Regis Hotel. We cemented our wedding vows at the charming neighborhood church of Saint Thomas Moore. The reception and luncheon were held in lower Manhattan at the River Club. My ushers were my brothers, Mark and Booker, along with David Davenport and my college roommates Gordon Lyle and Morris Gray. Jeanne balanced this group with five elegantly dressed bridesmaids. The reception went on until we were poured on the airplane for Albuquerque. We finally arrived at two o'clock in the morning, spending what was left of us and the night at the Alvarado Hotel.

I had a car stashed at the hotel and the next day we started on our honeymoon trip. This was a short journey, as our car was hit by a garbage truck, damaging the headlights. So, after the party at the fancy River Club in New York, we spent our first real married night at the Ace Motel in charming Truth or Consequences, New Mexico.

We arrived at the La Jolla Beach and Tennis Club on December 1 and honeymooned there until December 14. We then drove to Los Angeles. Jeanne had lived in LA before entering the Navy and I had some business contacts to make. She saw her friends while I inspected some supermarkets and visited with several canneries. We drove back to New Mexico via the Navajo Reservation and arrived at our new home in Santa Fe on January 1. We were awakened that night by a noisy chivaree orchestrated by my sisters and our invading friends the Brosseaus, the Eddys, the Greers, the Lords, and my two brothers.

That same day, I had received a call from Mr. Scott, general manager of the Kimball Company. Mr. Scott greeted me and went on to say Mr. Kimball would like to buy Gross Kelly & Co., and that if we didn't sell, Kimball would proceed to break us. Kay Kimball was the president of Kimball Grocery Co., whose headquarters were in Fort Worth, Texas. Kimball was a successful venture capitalist. In addition to wholesale grocery operations in West Texas and eastern New Mexico, he had extensive oil producing properties and food processing businesses and he was an avid and knowledgeable art collector. Competition was his love and his vast assets made him a formidable competitor. The call was indeed a momentous end to a tranquil honeymoon and the start of a turbulent business career and an eventful marriage.

1954: A Search for a Career

Jeanne and I settled in a comfortable rental unit at 355 Palace Avenue—remarkably, only a five-minute walk from the parental homestead at 531 Palace. Lois Field, an old family friend, had bought the gabled Francisco Hinijos house as an investment and to guarantee its preservation. She had also furnished it with some of her antique furniture, which harmonized with the unique house. It had a charming parlor with dormer windows, which looked out on a broad porch lined by a white, brocaded wood fence. There was a dining room, a kitchen, a large master bedroom, a bath, and a small alcove nestled under the stairwell, which led up to a large, empty attic. Jeanne quickly transformed the house into a comfortable and attractive home. The pace of our lives was rapid and exciting. Jeanne became a key player in the activities we and our friends were involved in, both professionally and pleasurably.

I concentrated on trying to manage the business while at the same time negotiating with Kimball Grocery Co., which was intent on purchasing Gross Kelly & Co. The Gross family stockholders, all living in St. Louis, were anxious to turn their stock into cash. Kimball was an awesome competitor and our business was becoming less and less profitable. Times were changing and our organization was aging and under-capitalized. At a stockholders meeting held in Las Vegas, New Mexico, on April 28, the vast majority of the stockholders voted to accept Kimball's purchase offer. The die was cast and I became totally absorbed in completing the transaction as profitably and as smoothly as possible.

Since Kimball did not purchase the Gross Kelly company stock—his company bought only its assets, subject to audit—my task was to maximize the stock value and liquidate the company for the best benefit to the stockholders. Inventory, including furniture, fixtures, and vehicles, was taken at each branch during May and June. We also evaluated accounts receivable. For the balance of the year, I stayed busy collecting charged-off accounts and paying liquidating dividends to the stockholders. Our general offices moved from Albuquerque to our cozy rental home at 355 Palace Avenue. Jeanne became my secretary.

The activities of our busy personal lives outside Gross Kelly & Co. did not stop. Jeanne was pregnant and expecting in September. I began looking for a job to support my wife and coming family. Dad invited me to the Bohemian Grove encampment in California in July. There, I spent ten days in the company of many of the important men in the country, absorbing their wisdom and consuming their liquor. One of the highlights of the visit was an eightieth birthday party for President Herbert Hoover.

In August, Jeanne's mother, Betty Wise, visited us. She was gregarious and a popular guest at our gatherings. An excellent pianist, she could pick up any melody on a moment's notice. She also was a sympathetic mother to her daughter, who was soon to be a mother herself.

On September 27, 1954, I took Jeanne to St. Vincent Hospital. After a very laborious delivery that culminated in a Cesarean section, a lovely little blond girl, Susan, arrived on the scene. I lived again like a bachelor, being fed by friends and family and tending to the many affairs and responsibilities that had now become a part of my life. I was grateful when the baby, Jeanne, and a nurse arrived home on October 6. A new page was inserted in our life's book. I became a parent, a role that continued to evolve with each new child and has really never ceased.

Along with the responsibility of liquidating Gross Kelly & Co., I was busy looking into job opportunities. The editor of *The New Mexican*, Lincoln O'Brien, offered me work, as did Vernon Wasson, president of the First National Bank of Santa Fe, who asked me to take over the bank's Los Alamos branch management. At the same time, my ranching friend, Malcolm Stewart, was looking for an infusion of cash for his operation in southern Colorado and asked if I'd like to join him. I would have liked getting involved with ranching again, but I did not have the capital he required. John Dillon, the new owner of the Santa Fe Book & Stationery Company, was anxious to share his business problems with a partner and approached me to see if I'd be interested, but maintaining our friendship was more important to both of us than co-owning a business. Also during this busy time of impending transition, I was involved in prospecting for uranium deposits along with my friend Newt Eddy, when time permitted. We combed remote areas of the Four Corners region and the mesas of southern Colorado with our Geiger counters. We covered a considerable expanse of country, but nothing activated our instruments.

In November, Jeanne and Susan were ready to take a trip. Jeanne wanted to show off the new baby and take a break from me. They flew east and joined Jeanne's parents, who commuted biweekly between their lovely farm on the eastern shore of Virginia and an apartment in New York City.

I flew to New York to join Jeanne and Susan and to see my in-laws and look into job opportunities. I visited the Bankers Trust, the Borax Co., Grace Line, the Guarantee, and Chase Bank. My contacts at these businesses generally received me warmly, but Jeanne was not in favor of moving back east. She felt that if I took a job with any of these businesses, I would be an unwilling western captive of eastern capitalism.

We departed from La Guardia Airport on December 14. I vividly remember saying goodbye to Harold, Jeanne's father, and watching him give his granddaughter Susan a warm goodbye kiss. Just a week later, on December 21, 1954, Harold Wise would be dead.

Although Harold had retired, he had been recalled as chairman of the board of McFadden Publications. His monthly work schedule alternated between Virginia, where he was a gentleman farmer, and New York City, where he served as the company's chairman. He had been entertaining business guests at his New York apartment when one of them inadvertently dropped a lighted cigarette into an upholstered chair. Waking late that night, Harold endeavored to open a window but was asphyxiated by the smoke that had filled the apartment. This charming gentleman was only fifty-eight years old.

Jeanne and I immediately flew back to New York, leaving Susan with her grandmother Ranny. It was a grim and tragic Christmas. Harold Wise was buried in Kensico Cemetery in Westchester County, New York, on Christmas Eve, and we returned to New York City on Christmas Day. I returned to Santa Fe on December 27 and Jeanne accompanied her mother back to Virginia. There she ended this benchmark year of 1954, comforting her grieving mother.

1955: A Full Life

In January Jeanne went back to her demanding career as wife and new mother in Santa Fe. Life was a blur as I juggled domesticity, uranium prospecting, job hunting, liquidating Gross Kelly & Co., and a stint in the vending machine business.

Domestically, we kept the extracurricular calendar full with parties, skiing, trips around the state, golf, fishing, and Fiesta. Uranium exploration and stock promotional deals kept western New Mexico and southern Colorado buzzing with real and false uranium discoveries, drilling and mine developments. Newly minted companies promised overnight riches. I was active with Frontier Uranium, Inc., located in the Gallup area, where discoveries of ore deposits led to stock issues and sales and, later, to total deflation, when the promotional ore discoveries turned out to be simply "rat holes" with no real ore deposits.

Oil and natural gas exploration also was active in northwestern New Mexico and eastern Colorado. I became involved in this activity through my brother-in-law, Frank Gorham. An able petroleum geologist, he was the president of Pubco Petroleum Co., a subsidiary of the Public Service Company of New Mexico. His focus of interest was in the San Juan Basin in northwestern New Mexico, where I went on several field trips with him and learned the basics of surface geology. Frank was an excellent teacher and a smart businessman; he wanted no relatives in his company. He pointed me north to Denver, where I interviewed with Shell, Sinclair, and British America, but received no job offers.

In Santa Fe I negotiated with Ken Pike, who owned a Redi-Mix concrete business. He wanted to sell his business, but the selling price was out of my range. Yet another business opportunity arose when Arthur Bosworth, a friend of my father's in Denver, suggested I open a branch of his stock brokerage business in Santa Fe. At the time, I did not feel Santa Fe offered a large enough market potential to make this venture profitable. Later developments proved me wrong.

As far as liquidating Gross Kelly & Co. was concerned, I continued recovering previously charged-off accounts and working with creditor committees endeavoring to salvage failing grocery stores, which owed the company money. I was quite successful in this work and, in time, recovered over 104 percent of previously charged-off accounts. (I did this by picking up payments for accounts that weren't on the books anymore). This did not increase my popularity with former creditors in the business community.

In September my luck with job hunting changed when two life insurance companies, New York Life and Penn Mutual Life Insurance Co., tried to recruit me as an agent. I knew nothing about life insurance, but was at the point where I needed to fix on a permanent career —Jeanne was pregnant again. I sought the counsel of Howell Earnest, who owned and operated a successful general property and casualty insurance agency in Santa Fe. Shortly after this meeting, Howell's business partner suffered a heart attack while playing golf and died on the golf course. Howell contacted me and offered me a job in his agency. His offer was more than generous. I was to work as a salaried employee for a year. If our relationship worked out satisfactorily, I could option to buy a third of the agency, using earnings derived from the business. Furthermore, I would have a five-hundred-dollar monthly expense allowance and he would send me to an insurance company training school in New York City for six weeks. I was not sure I would like the insurance business, but, encouraged by Jeanne, I accepted his offer.

Howell Earnest was about fifty when he and I joined forces. His wife was a puckish and humorous woman with a large and friendly heart. She had come to Santa Fe as a trained teacher for the deaf. Here she met and married Howell. They soon became known as a receptive and friendly couple and were well liked in Santa Fe. They boasted about and adored the little girl they adopted from St. Vincent's orphanage. Howell's father, from Clayton, New Mexico, had come to Santa Fe as an administrative officer in the state government. Later, in 1926, he founded the insurance agency that bore his son's name: Howell Earnest Agency. Howell attended the Los Alamos Ranch School for five years, graduating from this excellent institution in the late 1920s. The untimely death of his father thwarted Howell's college career, requiring him to assume the operation of the agency.

By 1955 the agency was in its thirtieth year and was a professionally operated and profitable business. It represented prime insurance companies such as Chubb, St. Paul, Home and other firms. It furnished the insurance needs of many businesses as well as a large noncommercial clientele. Howell was a bank director and also served pro bono on the board of several organizations. When I joined his agency, he was the chairman of the New Mexico Penitentiary Board, a selfless and demanding volunteer position by appointment of the governor of the state.

Upon accepting his job offer, I immediately enrolled in the insurance course offered by the Royal Insurance Co., a large British firm with U.S. general offices in New York City. On September 28 Jeanne, Susan and I headed for Virginia, to her family's farm. This would be Jeanne and Susan's home for the next two months. After they were settled in, I left and took up residence at the Harvard Club of New York City, where I had previously become a nonresident member.

Our temporary living arrangements at the Wise farm were scarcely a hardship. The farm, called the 5-Y, was located on the eastern shore of Virginia, on the tail end of the Delmarva Peninsula. The town nearest the farm was Onancock, founded in colonial times just across the Chesapeake Bay from Norfolk, Virginia. The farm was a comfortable operation run by a genial and able black foreman, Henry Wise, whose last name was coincidentally the same as his employer's. The farm lay on a neck of land jutting out into the bay. Beautifully manicured fields of soybeans, corn, pasture and woodlands surrounded the comfortable manor house. Fishing, duck hunting and, for the refugee from New York City, golf, were popular attractions. Foreman Wise's large family afforded an ample supply of cooks, maids, and baby sitters. I loved being there on the weekends, free from the insurance school.

My living situation in the city was also anything but painful. I had a comfortable bedroom. The Harvard Club's dining room offered excellent meals and its squash courts were great places to limber up after eight hours of classroom studies.

The course of studies at the insurance school repeated much of the business administration subjects I studied at Harvard. I was impressed to find the insurance industry played such an integral and important role in the business life of the country and the world. This made my decision to participate in the industry much more palatable than I originally anticipated. I realized that, at last, my MBA degree would serve me in making my new career a success.

The Royal Insurance Co. school was located in the Wall Street area. I became a commuter, taking the subway from Times Square to Wall Street. I experienced this financial hub of the world for eight hours each day for most of two months. At the end of each day, I usually retraced my route to the Harvard Club, where I often played squash or enjoyed some other diversion before dinner. Then came the evening studies in preparation for the next day of classes.

On weekends, like a typical New Yorker, I swerved from the hectic New York speedway to the tranquility of rural Virginia. In addition to the training, these two months exposed me to the lifestyle of a commuter to and from the business heart of New York City. I realized ever more clearly that uncomplicated New Mexico had its pluses.

The insurance course ended on November 21. I joined my family in Virginia and we boarded the plane for New Mexico. On our second wedding anniversary, November 28, 1955, I commenced my first day as an employee of the Howell Earnest Insurance Agency.

1956: Arrival of the Twins

Once I was in the insurance business in Santa Fe, I became deeply involved in prospecting for new accounts. I knew many potential clients and, although I was new and learning, my general business training and experience with Gross Kelly & Co.'s operations opened many doors. In addition to seeking out new business, I also serviced the agency's large number of clients, reviewing and updating accounts and handling claims.

To be successful in business, maintaining high visibility and pro bono involvement in the community is essential. As I sought out business, pro bono activities were ultimately self-serving, but also were part of the strong civic service and social responsibility syndrome that existed in the Santa Fe business community at the time. I was a member of the Rotary Club, which met weekly at La Fonda Hotel; the organization was always involved in some form of social betterment activity that required my time. I also joined the Junipero Serra Club, a lay church organization that assisted young men, financially and socially, during their studies for the priesthood at the local seminary. I was also one of the directors of the Boys Club, which always required funds to keep the facility intact and the sport activities going. Additionally, I became chairman of our local voting precinct, which required attendance at and direction of precinct meetings.

By 1956 skiing had become a great attraction and an important contributor to Santa Fe's economy. The ski area and its lifts and facilities had moved to the flank of Lake Peak, having abandoned the primitive, original ski run at Hyde Park. The Santa Fe Winter Sports Club, of which I was an active member, promoted races, activities for children, and social events at the ski area. We also organized promotional efforts in Santa Fe. I, along with other club members, served on the ski patrol, assisting management in maintaining discipline on the slopes, and rendering first aid and evacuation of injured skiers.

The newly acquired "profession" of being a husband and father also took up much of my time, of course. In this, I was not alone. Like me, most of my friends had returned from the service and had been or were soon to be married. All of us were young, our wives were socially active and most of them were young mothers with children of a similar age. The demands on all of us new husbands and fathers were similar. Tending to finances, child caring, homemaking and social diversion were an extremely high, time-consuming priority.

All of these activities left very little time for hunting, fishing, golf, tennis, skiing, and hiking. Yet these were the potions that kept our mental and physical systems in balance. Somehow, we fit these remedial pursuits into our somewhat frantic schedules.

Just when I thought I had it all more or less under control, Jeanne delivered twin boys. The arrival of Robert and Tom on April 16, 1956, dethroned Susan, who was only a year and a half old. They disrupted our domestic routine completely; suddenly we had three children under three years old, in a house with one bedroom and an alcove under a stairwell, which was to serve as the boys' dormitory. Jeanne's mother, Betty, arrived to try to bring some order into this chaos. She helped, but understandably did not extend her visit. On May 5 the boys were baptized. Becoming members of the church did not help. They continued to throw up almost immediately after being fed. Thankfully, conditions gradually improved and married life settled into a demanding but organized routine, which most of the time was pleasant.

Several events took place later in the year that added spice to our fare.

One of the highlights was attending the gala production of *Tosca* at the quaint opera house at Central City, Colorado. It was preceded by a black-tie, pre-opera tailgate supper. Joining us in this party were Malcolm and Mary Helen Stewart, David Wilhelm and his wife, and Jack and Betty Malo.

Malcolm operated three large ranches in southern Colorado. David ran the Wilhelm Feedlots Company and Jack, grandson of the milling king J. K. Mullin, manufactured livestock feed used in growing and fattening cattle. These three managed an integrated beef production syndicate. I had initially introduced the three to each other. Malcolm had been an artillery forward observer in the Tenth Mountain Division. David was an ex-fighter pilot and an eight-kill ace in Europe. Jack was a former executive officer of a naval destroyer.

It was a fine gathering. *Tosca* was Jeanne's favorite opera and I was with three good friends, all of them involved in my favorite avocation, the cattle business.

A quite different event, and a sobering education, was a tour of the recently completed New Mexico Penitentiary complex. Howell Earnest was on the penitentiary board during a renovation of the facility. He invited Jeanne and me to the formal dedication ceremony and to a tour of the facility. Seeing the cell blocks, solitary confinement area, a new model electric chair, and the inmates massed in the free-time center gave me a graphic insight into the complex effort to deal with a dysfunctional segment of society.

Almost monthly, I visited in El Paso, Texas, one of our important insurance markets and the site of the regional office of the Employers Group, operated by Echlin, Irvin and Crowell, an established general agency. A business day at their office was followed invariably by *una noche* on the town in Juárez. The Echlin firm and our agency were closely related and the partners in the firm always gave us an extra dose of nightlife across the border.

Christmas 1956 was mostly for Susan. The boys were still too new to the world, but Newt Eddy, dressed as Santa Claus, appeared and the jolly one set the stage for the coming year, which was to be a year of settling in and establishing the foundations for growth in my business career and in our family life.

1957: Foundations for Growth

In January 1957 I officially became a partner in the Howell Earnest Agency. I was given a substantial minority interest in the business, predicated on buying it over a ten-year span. In this way, I could benefit from the earnings of the company and use my portion to make the annual repayment. The business prospered and I found the annual payment a manageable financial hardship.

As my family grew, the housing squeeze was becoming more severe. The alcove where the boys slept was scarcely worthy of the occupation. Over the three years we had lived in the Hinijos house, we had grown in number from two to five and the youngsters were increasing in size. We needed more room, which demanded additional money to acquire the needed space. My grandmother's house, next door to my family's home, had been rented out since her death in 1952. The rental payment was helpful, but tenant turnover had been frequent, probably because of the numerous Kellys and their perpetual motion next door. Dad was looking for a permanent tenant or owner for the property. His constant patience and generosity towards his children directed him to consider our housing needs. He offered the house to us on terms that would have been hard to replicate elsewhere. Living next door to one's mother-in-law was not recommended, so I left the decision on the offer up to Jeanne, cautioning her about the hazard. She evaluated the risks and the offer and gratefully accepted both.

On March 21, 1957, we moved from 355 to 535 Palace Avenue, which has been home for me ever since.

The "Kelly Compound" had an interesting history. Ashley Pond, the founder of Los Alamos Ranch School, bought the property at 531 Palace from Don Amado Chávez shortly after World War I. The spacious property included a rambling adobe residence on the west edge of the lot, while a garden took up most of the center and eastern portions. My parents bought the residence and about half the land from Pond in 1925. Pond built a new home and a tennis court on the eastern portion of the property and this became 535 Palace. The two properties were separated only by ownership. In 1934 my grandmother, who was widowed, bought 535 Palace from the Ponds and moved next door to her beloved son, my father. When she died, 535 became a rental until 1957, when my father offered it to Jeanne and me and it became our home.

The house was John Gaw Meem's first effort at residential design. It predated his Pueblo Mission style and was built more along the lines of a California hacienda. It was made, not of adobe, but of hollow brick tile. Its construction required excavation of part of a hillside. The house rose three stories in height, with a two-car garage and laundry area on the ground floor. Its four bedrooms, two-and-a-half baths, and large yard made it an ideal home for my family. One unnecessary appendage was an indoor fire pole dating from Ashley Pond's day as Santa Fe's fire chief.

With job and home requirements apparently filled, it was time to grow the business to pay for these assets. Howell's active involvement in political and financial areas helped the business. I concentrated on insuring the uranium mines, ski areas, general businesses, art galleries and homeowners. The agency enjoyed good growth. Pro bono activities consumed considerable time, both during and after normal business hours.

In November, Jeanne visited in New York City and Virginia, leaving me with the care of our young family. Needless to say, I imposed considerably on my long-suffering parents next door and on baby sitters to help me manage.

Upon Jeanne's return, my father presented me with a favorable three-year note, converting the rental payment to house equity payments, and in December 1957 he gave Jeanne and me a warranty deed to our house. To add to this Christmas gift, the agency presented me with a three-thousand-dollar bonus. With all of this, the year ended on a splendidly high note.

1958: Family Life

As 1958 dawned, a snowy winter deepened, keeping the Santa Fe Winter Sports Club very active. Several club members were on the ski patrol, which required us to be at the ski area most weekends to patrol the slopes, render first aid and assist in the frequent racing competitions. When we weren't on the ski mountain, we were attending club related meetings, trainings, and social events in town. The club was close knit, with most members of relatively the same age and marital status.

Our three children were four and two years old. Dee and Grace Lord's four children were all starting to ski, as were Dave and Effie Davenport's four children and Newt and Connie Eddy's three boys. We took on the chores of dressing, driving to the ski area, teaching and feeding the young mob. The mothers took turns supervising, freeing up other mothers to ski. At the end of a ski day, driving the tired, wet and noisy pack back to town often involved unsnarling stuck and stalled cars from snowdrifts and icy patches on the wintry road. Though exhausting, the family ski adventures were a time for family bonding and for distraction from work.

Living next door, my parents had their own full life. My ever active father attended board meetings of the museum, the bank and the Public Service Company. When not entertaining out-of-town guests, he waged almost daily games of canasta and dominos with any and all players he could recruit. Begrudgingly, my mother would help entertain his visiting guests when she was not involved in a study group, with the Altar Society, or seeking peace and solitude reading a good book.

At the same time, my parents were still involved with parenting my siblings Caroline, Mark, and Booker, who were living at their home. Caroline could not settle on a job or a potential husband and constantly dragged my parents into her emotional problems. Mark and Booker were attending law school in Albuquerque and required supplemental financing and housing when they were home on weekends. Furthermore, Jeanne would occasionally look to my parents next door for babysitting.

Skiing gave way when the late spring finally arrived. Outdoor activities with the children followed the change of season. I mapped out an elaborate vegetable garden and recruited the children to help with it, although it seemed they mostly got in the way and trampled on newly sprouting plants. Jeanne took charge of picnics and supervising the children and their friends who came over to play. I managed to squeeze in time for hikes and to play tennis, golf and fish.

On June 18 Booker underwent heart surgery at the Mayo Clinic in Rochester. Father attended the fiftieth anniversary of his class at Harvard June 8-12, and then met Booker at the clinic to accompany him to the hospital. The operation was successful, and my father stayed with Booker until June 30, when Booker was released and went to St. Louis for further convalescence. Dad returned home, but not necessarily to peace and quiet.

Jeanne's sister Betty Lou was engaged to marry Lawrence Oliver on July 19. To help with the arrangements, Jeanne joined her mother and the bride-to-be in New York on July 12, leaving me with the three children. During the day their saintly nanny, Mrs. Moya, ably cared for them. At night, it was my job. On June 18 I turned the children over to their grandparents and departed for New York to give the bride, Betty Lou, away. The groom accepted my gift at their marriage, and on July 22 Jeanne and I returned to Santa Fe to reunite with our temporarily orphaned children.

Senior partner Howell felt I needed to sharpen my skills by attending another insurance school, this time in Boston at the American Employers Insurance Company headquarters. The course was scheduled to run October 5 to November 20. This was a long time to be away from home and family, but the prospect was enticing. Boston was not exactly a strange place for me, and fall in New England could be pleasant. Jeanne acquiesced and I commenced the course on October 6, staying at the Harvard Club of Boston.

Once again, I commuted to school. The course was given on Milk Street in downtown Boston, in the heart of the city's financial district. Our agency wrote considerable business with the Employers group, and though I was a student, the management there treated me very cordially. The curriculum covered the main aspects of the insurance industry. The instructors were excellent and expected the best effort from students.

On weekends I felt almost like I was back in college. I contacted several of my former classmates, attended two football games, and spent a weekend cycling on Nantucket Island. I even visited my old school, Portsmouth Priory, where I saw two former students, now monks, I had known: Father Thomas Van Winkle, my former roommate, and Father Aelred Wall, who with two other monks later established the Abiquiú Monastery of Christ in the Desert. I also had the pleasure of taking two of my favorite masters of twenty years before out to dinner in Newport—Dr. Lally, the historian, and W. Griffith Kelley, my English teacher.

All in all, the insurance course was valuable, and the return to Boston was a sentimental journey back into the past. When I returned home I found my wife still glad to see me. On November 28 we celebrated our fifth anniversary. My diary noted, "nice trout dinner and *vin blanc* – just *entre-nous*," in my somewhat poor French.

December arrived, and with it another snow season commenced. Christmas was becoming traditional with the children, and we went to Glorieta Baldy to collect greenery. Dad celebrated his seventy-second birthday on December 16. He was showered with nonessential gifts by his spawn and by the invited guests.

On December 22 my mother-in-law Betty Wise arrived, and on Christmas Eve the piñon tree was decorated and surrounded by presents awaiting the chaos that would follow church on Christmas Day. On December 31 another active year came to a close with a quiet evening at home for a change.

1959: Uranium and New Interests

On January 1, 1959, Governor John Burroughs was sworn in. As usual, I attended the ceremony, but this time I took the three children. This initiated a custom they had to endure until they were emancipated as adults.

In February I sold my vending business to Tony Fiorino for $125 per machine. This venture had commenced in 1954 when I was searching for some income-producing activity after the sale of Gross Kelly & Co. I had been enticed by an advertisement extolling the virtues of dispensing hot coffee from vending machines located at strategic gathering locations. I felt it would be a good investment and would require little supervision and maintenance. My brilliant marketing talents acquired at the Harvard Business School lured me into this pit of frustration.

I purchased five hot coffee dispensing machines and placed them at several locations, such as the Lensic Theater, Ned Wood's Santa Fe Motor Co. and the U.S. Post Office. Inserting a quarter coin into the slot activated all the phases of coffee brewing, ultimately dispensing hot coffee into a paper cup. The coffee could be modified by adding powdered cream and ground sugar.

Unfortunately, the machines never operated properly and required constant emergency repair. Having no technical ability, I had to hire Bob Hart of the Coca Cola Bottling Company to be the emergency engineer. Obviously, I was delighted to be rid of these recalcitrant, mechanical brew monsters when I finally sold them.

I celebrated my liberation from the vending business by taking Jeanne on a February birthday trip to Tucson and Guaymas, Mexico. In Tucson we stayed at Arthur Pack's Ghost Ranch Motel, named for his ranch of the same name at Abiquiú, New Mexico. While in Guaymas at the La Playa Hotel, we discovered Enpalma Beach some miles south, a gorgeous sandy stretch, miles long and occupied only by large assemblies of pelicans and cormorants. It later became a favorite beach for our Santa Fe Kiva Club members.

Upon returning to Santa Fe, I was busy doing insurance safety engineering for Sabre-Piñon, which operated several uranium mines in the Ambrosia Lake area near Grants, New Mexico. Dick Bokum, the president of Sabre-Piñon, had turned over this important insurance account to our agency. Finding a suitable insurance market for Sabre's uranium mines and its large trucking division required in-depth safety inspections by the insuring company's engineers.

In order to inspect the mines, we would fly from Santa Fe to a landing strip near the mines. From there, we would be driven to pit locations for inspections. Bob Dodge, Bokum's brother-in-law, was the company pilot, flying a Beechcraft.

One trip was particularly interesting. Bob and I took off from Santa Fe and picked up the safety engineer in Albuquerque. En route to the mine, somewhere above the majestic volcanic plug called Cabezón, the landing gear indicator commenced flashing, but we couldn't tell if the landing gear was down or up. We turned back to Albuquerque to get a reading from the airport's tower personnel. On the approach to the Albuquerque airport, though, our radio failed to transmit, so we could only listen to questions from the tower and couldn't reply. To solve the problem, Bob flew the plane tower-high along the length of the runway, wagging the plane's wings. The flight controller surmised we were concerned about the landing gear and transmitted the message, "the gear is down, but we can't tell whether it is locked down." The controller instructed us to make a landing and alerted ambulances and fire engines to stand by. Fortunately, we made a safe landing and made repairs. We proceeded to the landing strip at the mine, buzzed it to clear away the goats, and landed without incident. We flew the nervous engineer back to

Albuquerque, but then, en route to Santa Fe, I had to hold the cabin door shut in flight, as it hadn't been properly locked. It was a day to remember.

In March I was elected president of the Santa Fe Rotary Club, to take office in mid-June. In order to be properly groomed for this exalted position, I was sent to New York City for indoctrination at the International Rotary Convention, June 6-14. Most of the meetings were held at Madison Square Garden. Attendees came from all over the world and numbered around one thousand. I had an interesting time and returned to Santa Fe highly charged, poised and ready to lead the local club in its varied activities during my term of office.

(I remained a member of Rotary for over twenty-nine years and finally resigned when my daughter, Pamela, applied for a Rotary European Scholarship; applicants could not be related to anyone associated with Rotary.)

Mid-summer 1959 also saw the birth of the Kiva Squash Racquet Club. A number of men in Santa Fe who had played squash at college organized the club, fired up by the interest of Harvey Durand, owner of La Posada Inn. Harvey converted one of La Posada's guest suites into a singles court, and the Kiva Club was born. I became an active member and remain so today.

In addition to the facilities in Santa Fe, the Kiva Club maintained seaside lodging at San Carlos beach near Guaymas, Mexico. Club members could rent the facility and take their families to the seaside resort during the cold winters in Santa Fe. Enpalma Beach was one of the popular attractions there. The drive from Santa Fe to San Carlos took two days. Keeping the children from quarreling with each other on the long drive was usually an insurmountable challenge. However, the arrival at the beach made the ordeal well worth the while.

The Shalako ceremony at Zuni pueblo took place on December 5, just before the anniversary of the Pearl Harbor bombing. I took the opportunity to attend the Shalako and to make an insurance check on Sabre Pinon's uranium mining activities in the Grants area. With friends, Jeanne and I took in the all-night Shalako House Blessing ceremony and the Mud Head ritual. The next day, Sunday, we visited Window Rock, the capital of the Navajo Nation. There we went to Mass, which was a pretty dull affair compared to the Shalako and Mud Head rituals the night before. On our way back to Santa Fe, I checked on Sabre Pinon's mine at Ambrosia Lake and the company's trucking division at Grants.

Christmas was exciting for the Kelly children. Ed Brosseau took on the role of Santa Claus for them, questioning them on their behavior. Being jolly and forgiving, he was more than generous in his gifting.

The year's last entry in my diary reads:

"All in all another good year and an eventful decade– healthy and happy—very much to be thankful for—may the '60s be the same! Adieu."

1960-1969

THE SIXTIES: LEARNING, ADJUSTING AND ACHIEVING

1960: Business and Family

By 1960 we were pretty well settled in to our life in Santa Fe. We had been married for seven years and had produced three children. I was in my sixth year as an insurance agent and I had become seasoned and reasonably well known in the business and civic communities of Santa Fe.

From a business point of view, the influx of residents after World War II brought a new crop of entrepreneurs into Santa Fe. Existing businesses were invigorated by the employment of these newly arrived citizens. Furthermore, these new citizens, who were charmed by the cultural diversity and recreational opportunities of the area, brought fresh ideas and enthusiasm into the social life of the community.

My close friends were part of this infusion. Dick Bokum from Chicago became an active political player, a building developer, and a uranium producing executive. Andre Senutovitch prospected and developed several uranium mines in the Grants area. Leland Thompson became the president of the Western Development mining venture. Newt Eddy and Edward Brosseau were active in prospecting and drilling ventures. Nathan Greer took over his family cinema and real estate company and modernized and profitably expanded the empire. John Dillon bought the Santa Fe Book & Stationery Co. and housed it in a new building, owned by Newt Eddy. Carl Goodwin operated the two radio stations broadcasting in the region. Nash Hancock and

Tom Old established a new automobile garage and car dealership. Thomas Catron became the third generation member of the Catron law firm. Don Van Soelen was the popular new banking personality at the First National Bank.

Santa Fe became a ski destination, drawing people from other states to come and ski. The Santa Fe Winter Sports Club became the home of the local ski patrol. The club built and operated the Butler recreational center. I was elected head of the Ski Patrol in 1960. This honor required I participate in patrol duties on almost all winter weekends.

The United Fund of Santa Fe had become the funding mechanism for the social welfare organizations of the city. I had been involved with raising funds to finance the local organizations and was elected president of the United Fund in 1960. My fundraising efforts would have been next to impossible without Bill Hooton's active management as the director of the United Fund.

In spite of the competition for my time from these extra-curricular activities, marriage and family life retained priority. In February Jeanne and I took a vacation in Aspen, Colorado. Perhaps as a result of skiing moguls, she became pregnant. On September 26 Jeanne produced little Pamela, our fourth and last child, by another Caesarean section. Five years younger than her nearest sibling, Pamela became our pet.

During the year, Castro and Cuba were having a love affair with Russia, which aroused much anxiety in the United States. With the increasing tension in geopolitics, I was afraid I might be recalled into the army. In retrospect, I realize an officer with four children was more responsibility than even the military would care to assume.

In August Bob Lockwood remodeled parts of our home, finishing the job in time for Pamela's birth. We would start this new decade of the sixties with a new daughter, a remodeled house, and much good fortune.

1961: Community Involvement

Governor Edwin Mechem and President John F. Kennedy both took their inaugural oaths on January 1, 1961. Kennedy's memorable speech, in which he asked the citizenry to accept challenges and to make sacrifices, electrified and at the same time worried the nation.

The Russians seemed to be outshining the United States on the world stage. On April 12 they launched the first man into space. For our country, this was like suddenly losing a long-held world championship. At the same time, the U.S.S.R was converting Cuba into a rocket launching platform that would threaten the U.S. mainland. In response, Kennedy authorized a CIA-backed invasion of Cuba by anti-Castro factions. By May 17 the anti-Castro forces, and by extension the United States, were humiliated by complete defeat. The debacle fortified Castro's power, and Russian influence seemed to be taking over in Latin America. With Russia and Cuba's overt aggression, the possibility of nuclear attack festered like an ulcer beneath the routine of every-day life.

St. John's College of Annapolis decided to establish a second campus in Santa Fe. John Meem's generous offer of some two hundred acres of land—and the persuasion of Robert McKinney and others—had clinched the decision. I saw this development not only as a business opportunity, but also as an exciting addition to the cultural life of Santa Fe. I was privileged to be at the first meeting of the St. John's Board of Regents along with the provisional group who made up the Santa Fe welcoming committee. From then on, until the start of the first academic year at the new campus, St. John's became another pro bono project for me, as I participated in fundraising, construction, and public relations efforts on behalf of the school.

I also became involved with the Council on Crime and Delinquency, a national organization that added New Mexico to its area of activity because of the high degree of youth crime and delinquency in the state. I was asked to be on the council's state board, which met monthly and also held meetings and inspections at troubled facilities throughout the state. This was a real learning experience for me, as I became familiar with the operation, administration and challenges facing several state-run institutions and organizations that dealt with juvenile crime.

During a trip to inspect the boys reformatory at Springer (the New Mexico Industrial School for Boys), I was reminded of a remark my father made to me when I was twelve years old: "I have been thinking of sending you to a proper institution, and I am debating between the boys' reformatory school at Springer, New Mexico, and Portsmouth Priory prep school." He picked the one with the longest sentence: five years of boarding at a Benedictine monastery in Rhode Island.

Several new developments spiced up the day-to-day routine of the insurance business. In April, Bob Weil moved to Santa Fe and with his high energy and good business sense began new ventures. I was fortunate to meet Bob soon after his arrival and to serve as his insurance agent over the subsequent years. One of Bob's first moves was to buy the Bond Ranch, which stretched from near Santa Fe to Buckman crossing on the Rio Grande. Part of the ranch was on land leased from the state. Bob made the winning bid on this sizeable chunk of land when it came up for auction. He and his wife Zanie later created the imaginative and tasteful development of La Tierra on the ranch.

Concurrent with the ranching activity, Bob developed a cattle feed lot in the Moriarty area and utilized the King brothers' verdant corn and milo production to fatten local range cattle. Bob took flying lessons, became a bush pilot, and commuted from his ranch north of Santa Fe to his cattle operations throughout the region. Underwriting his business was especially interesting for me because of my background in ranching. We developed a close relationship.

John Dillon, another insurance client, was an elegant product of New York, and a squash-playing graduate of Yale. He had served in heavy combat action in World War II as an officer in Europe and was recalled for Korean duty. He and his wife headquartered in Santa Fe as he spent his Korean tour at Camp Carson in Colorado Springs. They fell in love with Santa Fe and moved here after completing his army tour. John became an entrepreneur and invested in several local retail operations as well as uranium mining speculation. He was also one of the founders of the now locally famous squash facility, the Kiva Club.

John and I became close friends. John bought from Dick Healy the Santa Fe Book & Stationery Co., the prime enterprise of its kind in Santa Fe. He later moved the business into a splendid new facility built by Newt Eddy expressly to house John's company. John was an idea man who delegated operational responsibility to others. In the course of time, the business seemed to be functioning well, but the profit margin was not in keeping with the sales activity. I had experienced a similar situation in my Gross Kelly & Co. days. The cause of the problem seemed similar, too: employee defalcation (embezzling of assets).

Since employee dishonesty was an insured risk, my company initiated an investigation, during which key employees were subjected to lie detector examination. The examinees were smart, though, and they knew if they took certain kinds of sedatives prior to the exam, the results could be legally non-conclusive. They did just that, and even though the finger pointed to inventory leakage, no conclusive evidence came from the lie detection exams. The suspected employees soon resigned or were fired, but the damage was done. John sold the business at a severe discount. So much for delegation of responsibility. (John had more success in his Washington Avenue liquor store, where he built upon his knowledge of wines and his partiality for gourmet foods. I insured this prosperous business and learned something of the wine industry as a result.)

On the night of June 23, the facilities at one of our agency accounts, the Pot Creek Logging and Lumber Co., a sawmill and lumber storage facility east of Taos, burned to the ground. No fire department service was available. Although stringent protective measures had been in place to prevent a fire. the mill and its lumber inventory were destroyed, and we paid a settlement of $120,000, which at the time was substantial. This loss affected the profit picture for what otherwise was a good business year for my agency.

Another extraordinary event took place in 1961: the brutal murder of Katherine Kavanaugh on September 28. Katherine was a bright and attractive widow who owned and operated the Los Pinos Guest Ranch at Cowles, New Mexico, adjacent to the Pecos Wilderness. She had inherited the popular guest facility from her father, Don Amado Chávez (the original owner of the house next door to ours on Palace Avenue). Katherine was an avid fisherwoman and an excellent horse person, as well as a charming hostess for this small, exclusive guest ranch. Katherine had hosted me at the ranch, gratis, in my younger years during my annual rides from Santa Fe to Las Vegas. Katherine would put my brother Hank and me up for the night, feed our horses, and serve us an excellent dinner. It was on one of these occasions that I met Frank and Robert Oppenheimer, later of Los Alamos fame. They "batched it" at their Pecos cabin and would ride horseback to Katherine's for a hot bath and for her charming company.

During the winter, "off-dude" season, when I was at the Harvard Business School, Katherine served as the dean of Finch Junior College in New York City. She was responsible for student social activities and organized dances for the girls. She looked to me several times to furnish suitable young men to attend their affairs.

Katherine was at home alone in Santa Fe on the night of September 28 when a burglar invaded her house. She surprised the culprit and he grabbed a kitchen knife and savagely killed her as she fought for her life. The murderer was apprehended and sent to prison, which seemed a small penalty for such a heinous crime. Los Pinos was operated for a time by her nephew, Amado Chavez, and currently continues under different ownership. (Incidentally, Lake Katherine, in the Pecos Wilderness, is named after Katherine.)

On a happier note, in November my father added another arrow to his quiver of clubs and orders when he was inducted into the ancient order of the Knights of Malta. This organization of eminent Catholic laymen was an outgrowth of the military-religious order that protected Malta during the Crusades. Dad was invested at St. Patrick Cathedral in New York City. Now, as Sir Daniel, he had an excuse to visit Rome and attend the ceremonies at the order's ancient headquarters, adjacent to the Vatican.

My diary noted that on December 20, 1961, I took the children skiing and they were "skiing quite well."

1962: Playing Cowboy

January 6, 1962, marked the fiftieth anniversary of New Mexico's statehood. The auspicious occasion was celebrated at a gala birthday party at La Fonda. Governor Mechem served as the titular host at this, the last of the several fetes celebrating the New Year.

Concurrently, several Santa Fe citizens were exploring the feasibility of organizing a private, non-sectarian college preparatory school in the community. Los Alamos Ranch School and Brownmore Girls School had been successful, but they were orphaned during the war years. It seemed to some community leaders that a comparable educational facility would serve Santa Fe well. An exploratory committee headed by Reverend Seaman, the Episcopal rector, formed in 1961 and by 1962 had decided to start such a school. The initial committee included Reverend Seaman, Dr. Ned Goodrich, Jim Adler, Carrington Wooley, Oliver Seth, and others. Bill Hooton and I became committee members in 1962 and were assigned the mission of finding a suitable headmaster to run the school.

On March 1 Hooton and I went to New York City to attend the annual convention of the Association of Private Schools. The convention was a kind of academic trade fair and a recruiting arena for teachers and potential headmasters. We immersed ourselves in the convention and after numerous interviews selected Doug McClure, then a master at Pomfret School in

Connecticut, as the best candidate for headmaster of the new school. Doug seemed adventurous, self-reliant, well qualified, and excited about being the head of a preparatory school in far-off New Mexico. We recommended McClure to the committee when we returned to Santa Fe and, soon after, he visited and was approved. He accepted the challenge.

In the meantime, the committee selected the unoccupied Breese Burner lab and office building on Upper Canyon Road for the school building. The lab, situated on the Breese Ranch, had been mothballed after the war ended, and needed modifications. The committee also had to select faculty members and confirm student enrollment. All of this required money. Fundraising became a high priority for the balance of the year as we aimed for a target date for opening in the fall of 1963.

Our home front was erupting along with all the activity around the creation of the new school. Our son Thomas, one of the twins, constantly seemed to be challenging the domestic status quo. He tended to be the dominant twin and would assert himself in annoying ways to make his supremacy evident. His antics were amusing to his siblings, but he often tested my self control and parental wisdom.

One day Thomas boasted to his older sister Susan that he could jump from his third-story bedroom window down to a small outside balcony jutting out from the second floor, about twelve feet below. Should he miss the target, he would fall another twelve feet onto the cement entrance to the garage on the ground floor. Susan called his bluff, and he made the jump—successfully. She proceeded to rat on him to his mother, who demanded I take remedial action to put a stop to this kind of risky behavior. I cannot recall what my remedy was, but it made him well aware of what damage he could have done to his body had he missed the target.

Thomas's mischievousness was diverted shortly after that incident by our acquisition of Cabrito, a little orphaned billy goat we acquired while flying kites west of Santa Fe on Easter Sunday. We discovered the abandoned kid while retrieving a downed kite. After searching the area unsuccessfully for his mother and other goats, we adopted him and took him home. He was raised on a diet of Pet brand evaporated milk he suckled from a Coke bottle fitted with a nipple. He grew quickly and eventually developed some grace. He became a household pet and thought he was a human.

Cabrito's life ended with a glorious yet tragic finale. We had loaned him to Tom Old, who had several nanny goats at his ranch house on Vivash Mesa east of Pecos. Tom savored roasted *cabritos* (baby goats), but he lacked a necessary element for producing them: a billy goat. So I loaned him Cabrito, our goat-turned-cavalier. Cabrito seduced the willing nannies but unwisely consumed some used crankcase oil and died from this ingestion. In due time, the nannies gave birth to several *cabritos*, our little Cabrito's offspring. Cabrito had done his job, and Tom Old had a good supply of cabritos.

In June the children were invited to their grandmother's coastal Virginia farm, where they spent a glorious month with her and Jeanne, away from Santa Fe and their short-tempered father. While they were gone, I had a chance to play cowboy.

My friend Malcolm Stewart owned a twenty-thousand-acre summer ranch on the Brazos River southeast of Chama, New Mexico. He had just taken delivery at his home ranch near Alamosa, Colorado, of 1,400 head of Mexican steers, and he needed to move these animals to summer pasture on the Brazos ranch. He invited me for the adventure.

These wild *corrientes* had to be shipped from Alamosa to Osier on the narrow-gauged railroad, the D&RGW, which ran from Alamosa west over the mountains to Chama. On June 15 we loaded the steers, about fifty to the car. At the end of the long train, a caboose and a livestock car accommodated the cowboys and horses needed to drive the cattle to the Brazos.

The train chugged out of Alamosa about four o'clock in the morning and proceeded to climb up to the station high in the mountains at Osier, about twenty-five miles from Alamosa and twenty miles shy of the Brazos pasture. When the train reached Osier, we released the Mexican low-desert cattle from the railroad cars. Thirsty from the long ride, they smelled the water from the Conejos River below and stampeded for the stream. After they drank their fill, we bunched the animals together into a trail herd and began the drive, following a long abandoned timber road to the Brazos river. After crossing several snowdrifts that still lingered along the route, the herd entered the grassy meadows of the Brazos ranch about sundown.

En route around noon we had arrived at a small, strongly fenced holding pasture where we corralled the herd and stopped for "chuck," the day's main meal. We were ravenous and greeted the camp cook and his creation with great anticipation. He prepared our meal in Dutch ovens suspended from an iron bar above a bed of hot coals. Beef stew, frijoles, biscuits and gravy and hot coffee made up the fare. It was delicious.

This meal reminded me of a vividly contrasting "dinner" I experienced years before while I was working on the Gross Kelly & Co. cattle ranch. We were near the end of a roundup, having been out for about a week. In those days, a roundup consisted of gathering the cattle from various sections of a ranch, sorting the cows and calves, and branding and vaccinating them. The mobile chuck wagon, where the cook set up his kitchen, was our home during roundups. Here the cowboys unloaded bedrolls from the wagon to spend the night.

Our camp on that night was miles from any habitation but just along the ranch's fenced boundary with the adjacent Bar-Y ranch. It was about five o'clock in the evening when the cook called "chuck time." The crew was loading tin plates with food when a Bar-Y line camp cowboy rode up to the fence. His isolated shack, or line camp, had not been resupplied and he was obviously half starved. We of course invited him for supper. He tied his horse, scrambled over the fence and joined us. When he saw and smelled the hot food, he could not contain himself. He exclaimed, "Jesus Christ, what a meal. All I have been eating the last few days is Mary Jane syrup and beans!"

My family reunited at the end of June, and the remainder of the summer sped by pleasantly, except for the growing and ever-present tension between the United States and the Soviet Union. In October I was hunting pheasants in South Dakota as the high international drama played out in the Cuban missile crisis. The duel between Kennedy and Khrushchev reached its dramatic climax in late October, when war seemed almost inevitable, but then the two countries finally reached a compromise. I, along with everyone else in the country, was greatly relieved.

U.S. Senator Dennis Chávez died in mid November. Governor Mechem resigned and appointed himself senator, to fill Chávez's seat. Lieutenant Governor Tom Bolack assumed the governorship—a position he held for all of thirty-two days.

As the year drew to a close, my duels with my son Thomas continued. He was the prime suspect when I found a thumb tack placed strategically on my dining room chair. I narrowly averted sitting on the tack. I arranged for Thomas to see a child psychologist. Upon completion of an evaluation, the psychologist was on Thomas's side and gave him a clean bill of health.

1963: New Mexico Amigos

Jack Campbell was inaugurated as governor on January 1, 1963. He would serve only one term and not seek reelection, although he was a popular and effective head of state.

Soon after the inauguration, I drove to Artesia, New Mexico, with Jim Gaskin, the accountant for the Wilson Oil Co. The purpose of the trip was to conduct an annual audit and to carry out an insurance inspection of the forty-some oil wells the company operated on the Merchant Ranch east of Artesia.

Parker Wilson, president of Wilson Oil Co. and his father, Francis C. Wilson, founder of the company, were close friends of my family and had lived in Santa Fe for many years. Francis, a Harvard Law graduate, came to New Mexico around 1900. Initially he worked in the legal department of the U.S. Indian Service in Santa Fe. He was actively involved in the legal affairs of both the Pueblo Indians and the Navajo Tribe. During the course of his career, he became a close friend of Chee Dodge, the chairman of the Navajo Tribal Council. Concurrently, Wilson became interested in acquiring oil leases in the Delaware Basin in southeastern New Mexico.

Wilson's financial resources were near exhaustion, prohibiting him from bidding on some potentially valuable leases. Word of his predicament percolated west to remote Navajo land, where Chee heard of Wilson's problem. Chee wired him to ask him to meet in Albuquerque at the Alvarado Hotel, adjacent to the railroad station. Wilson complied and met with Dodge, who offered him a thirty-thousand-dollar loan, which Wilson accepted. Subsequently, Wilson obtained the leases and developed the property, which became known as the Wilson Pool. It turned out to be a highly productive reefal oil field. (Incidentally, Chee's son later joined Wilson's law firm.) Parker succeeded his father in operating the oil business and was also active in other ventures, for which I was privileged to handle the insurance needs.

Upon returning from the Wilson oil patch, I found that Doug McClure, the headmaster of the soon-to-be-opened Santa Fe Preparatory School, had gotten cold feet and resigned. It seems the challenge of running a newborn school intimidated him. With the school primed to open in September, the need to find a suitable replacement head became a high priority.

Fortunately, Francis Bloodgood accepted the challenge and the job. He had just returned from running the U.S. Embassy School in Belgium and was on his way to a position with the Casady School in Oklahoma City when he was contacted by a Santa Fe Prep board member who had known him. The board approved of his hiring and he took on the job as the school's first headmaster, a position in which he served ably for four years.

The board tapped me to lead the school's first fundraising drive. Working closely with Bloodgood and members of the fundraising committee, we raised over one hundred thousand dollars by the time the school officially opened in September. The school was off to a great start.

I was also a member of one of the more successful public relations organizations in the state, the New Mexico Amigos. This group of men, active in business and politics, met annually to promote New Mexico. The drill involved a preparatory meeting, the fitting of colorful Amigos blazers on members, an organizational banquet, and a sojourn on a leased airplane.

For our 1963 trip, Governor Campbell was our leader. Among the Santa Fe representatives were Leland Thompson, Nate Greer, Tom Old, David Davenport, and I. We proceeded to Fort Worth, Texas, where civic leaders greeted us and showed us the Bell helicopter plant, the Six Flags Over Texas amusement park, and the General Motors assembly plant. Following the tour, we attended a banquet with Governor Campbell as toastmaster. Then it was on to Houston, where we inspected the NASA facility and took a boat trip on the Houston Ship Channel. After the Houston stop, we visited the Naval Air Station at Pensacola, Florida.

I had a particular interest in the air station because my wife Jeanne, a World War II WAVE, had been stationed there, where she instructed the flying cadets in free aerial gunnery. Our host for a tour of the station was Admiral Fitzhugh Lee, a relative of General Robert E. Lee. The admiral was an imposing figure and a classic southern gentleman. We boarded the aircraft carrier *Lexington* to observe landings and takeoffs by the cadets. When we visited the officers' wardroom, I noticed a bronze plaque honoring Lieutenant Laughlin Barker, USN, who during WW II made the one-hundred-thousandth landing on the carrier. I knew Laughlin well. He was an Annapolis graduate who served many years in the navy and returned to his birthplace of Santa Fe upon retiring.

From Pensacola our pilgrimage took us to the spacecraft launching area at Cape Canaveral in Florida. The trip ended up with a night on the town in the French Quarter of New Orleans. We returned to Santa Fe, steeped in aerospace technology and southern comfort relaxation.

Wherever we went, the New Mexico Amigos worked hard to promote New Mexico as a prime place to work and recreate. However, we were amateurs in the field of public relations and advertising compared to James R. Young, whose business I insured. Jim operated the only integrated apple growing and marketing facility on our insurance agency books. As a young man, Jim had sold sets of the *Encyclopedia Britannica*, an experience that led him to the giant advertising firm of J. Walter Thompson, which ultimately he headed. His son, Webb, lived in Santa Fe. On one of his visits to Santa Fe, Jim discovered a strikingly beautiful, serene and isolated valley north of the Cochiti Pueblo. He noticed an ancient, droughty, lonely apple tree in the middle of what once must have been a cleared field. He also observed the sunburned sides of this narrow valley and the green fingers of plant life following the intermittent stream that trickled down the valley floor. He was charmed by the discovery and intrigued by an idea.

Jim purchased the ten-thousand acre Cañada de Cochiti Grant, in the center of which lay this valley. Inspired by the ancient apple tree, he set about developing an apple orchard there. He hired a professional orchardist, Fred Dixon, who, starting from the one old tree, eventually developed a prosperous, one-hundred acre orchard, now famous as the Dixon Apple Orchard.

Dixon used sumps to bring an adequate supply of ground water to the surface for irrigation. With the rich volcanic soil, sunshine, and good fruit tree selection, he established a commercial crop within a few years. This was where I came in. Jim Young and I became friends and I became his insurance broker. Now that the orchard was maturing, it seemed prudent to insure the apple crop. I searched for an insurance market to underwrite the orchard.

The Pacific Northwest is a renowned center for apple orchards. Insuring the numerous orchards there against crop loss was not a problem, since insurance is based on the doctrine of spreading of multiple risks. The Northwest was home for many apple orchards and, thus, could be profitably insured. In Northern New Mexico, however, there was no way to spread the risks, since the only large commercial orchard was Dixon's. Nevertheless, I finally prevailed on the Insurance Company of North America to insure the Dixon orchard against, among other things, hail damage.

My company inspected the orchard, estimated the crop size, and decided on a premium and level of insurance. Jim questioned the high premium cost but paid for the policy. That growing season was favorable, and the crop ripened well, producing a large volume of apples with a high dollar value.

But just at picking time, clouds gathered, the temperature dropped, the mountain valley rumbled with thunder. A devastating hailstorm descended on this serene horn of plenty. When the storm wore itself out, over half of the crop was on the ground. The apples remaining on the trees bore the chickenpox-like evidence of hail damage.

My agency paid out forty-five thousand dollars for the orchard's loss, and I lost an insurance source for my client. But Jim Young, the eminent advertising marketer, salvaged the damaged crop by advertising and marketing "unique, hail chilled, mountain fresh, sparkling champagne, delicious apples."

The Dixon orchard remains in production, operated by Fred Dixon's granddaughter. Its apples sell out each year, but I doubt if the crop is insured. (The orchard was nearly destroyed by flooding after a devastating fire in the summer of 2011; its future remains uncertain.)

On November 22 I was in the Palace Bar in Santa Fe having lunch with my ex-Texas friend Leland Thompson. The bustling bar's TV was tracking President Kennedy's convoy in Dallas. Suddenly, the assassination took place and chaos ensued. I remember Leland's laconic comment, "Poor Dallas." The assassin Oswald was shot and killed by Ruby on November 24.

On a happier note, St. John's College in Santa Fe was formally dedicated on November 27. With the laying of its cornerstone, Santa Fe became the second home of the third oldest college in the United States.

When Christmas vacation started, Jeanne and I left Santa Fe with the children for the Sea of Cortez at Guaymas, Mexico. It was a striking change from wintry Santa Fe and a great place to end the year of 1963.

1964: Family Trauma

Fortified by a week at sea level, we left Guaymas and returned to Santa Fe, reasonably well prepared to face the new year. I was forty-three years old, had been married for ten years and was a devoted husband and conscientious father. The routine of our lives was hectic, but our activities were varied and certainly not boring. Things seemed to be falling in place nicely: business was good, community projects (such as Santa Fe Prep school and St. John's) were developing well, our family was healthy. Yet in some ways, I felt more like a spectator than an actor in my own life: as the action rushed by, I was observing, rather than participating.

Jeanne was a careful, competent skier, but her bones were fragile. On January 25 she fell and broke her leg. I became her nurse, as well as a nanny for the children. I soon realized this was a very demanding job. I did triple duty as business activities, housekeeping, and child nurturing took up most of my time through February and into March. However, I did enjoy being with the children, who by now could be helpful and, quite often, good company.

On March 28 a severe earthquake struck Juneau, Alaska. The news was shocking, especially since my close friend David Davenport and his family lived there. We were very concerned for their well-being. David's office building collapsed in the temblor, but fortunately, he was not in it at the time. His home, however, was shaken off its foundations and slipped precariously close to a cliff edge. He moved his family to Santa Fe while the displaced home was placed back on its foundations.

Several people who had had an impact on my life died that spring, including General Douglas MacArthur, the commander of U.S. forces during my tour in Korea, who died on April 5. Kay Kimball, the purchaser of Gross Kelly, died on April 15. Kay and I had become reasonably close friends, despite his ruthless business appetite. (His splendid art collection is held by the charming Kimball Art Museum, which he established in Fort Worth.) Only a few days later, on April 18, Herbert Clark, my father's sponsor at the Bohemian Club in San Francisco and a former stockholder of Gross Kelly & Co., died. Still another friend passed on that spring, Theodore van Soelen. "Soley," a close friend of my father, was an accomplished artist, a member of the American Academy of the Arts, a sportsman, and chairman of the State Police Board. Soley had painted portraits of my grandfather and my mother and father.

By the time June and the end of the school year rolled around, Jeanne's leg had healed and a change of scenery was in the offing. On June 5 we departed for the La Jolla Beach and Tennis Club in California. We crossed the Arizona and California deserts in June, which I don't recommend trying, especially with four children. Our route was south to Hatch, west on I-10 through Nogales and Yuma on the California border, and then on to the coast and charming and exciting La Jolla.

On August 4 President Johnson announced that the North Vietnamese had fired upon U.S. Navy vessels in international waters. In retaliation, the American ships sank a North Vietnamese boat. The event became known as the Gulf of Tonkin incident, and it foreshadowed full scale military action against the North Vietnamese—a chilling preliminary of things to come.

My father attended the annual gathering of the Bohemian Club in San Francisco. The two-week assembly at the Bohemian Grove was filled with prominent guests, as well as the regular members. Dad's diary mentions details of talks given in 1964 by General Dwight Eisenhower, German rocket scientist Werner von Braun, the well known filmmaker Lowell Thomas Jr., U.S. Attorney General Robert Kennedy, Navy Admiral Arleigh Burke, and U.S. Senator Barry Goldwater. I had been a guest at the Bohemian Club before, and this litany of great men was typical of the personalities one could see at the Grove.

At the end of the Bohemian Grove encampment, everyday life rudely jolted my father back to reality when my sister Caroline suffered a mental breakdown and was placed in the Nazareth Psychiatric Hospital in Albuquerque. Today, her problem would probably be diagnosed as bipolar disorder. This condition had surfaced from time to time in Caroline's life before, but this was a very serious episode. For the family, especially my parents, it was pure torture and extremely disrupting and frustrating, as well as expensive.

To further complicate my life, Booker's history of rheumatic fever triggered a severe heart condition that required open heart surgery. The operation, performed at Massachusetts General Hospital in Boston, was successful in repairing a damaged heart valve. In late December, as Booker was recuperating, Caroline's condition also seemed to be improving.

Jeanne and I became ensnared in all this trauma, but as sympathetic spectators rather than direct participants. As things settled down, my parents took a deserved week of rest and recuperation in Puerto Rico. They returned in time for a reasonably calm and unruffled Christmas dinner in Santa Fe, hosted by Jeanne and me.

1965: Acquiring the Agency

My sister Caroline continued in the role of leading actor in the Kelly family drama in early 1965 and for most of rest of the year. The cause of her neurosis was never pinpointed, but the condition was always lurking near the surface and could burst out suddenly and violently. She was a classic Dr. Jekyll-and-Mr.-Hyde personality. Over the years her condition would follow the same pattern: long periods of normalcy would be followed by periods of deep depression, resentment, self-pity, and indecision. Dramatic tirades followed. When her hostile moods descended on her, she would vent on family members, especially our father. She blamed him for her distress and made him the direct object of her anger. Caroline's outbursts were extremely distressing to Dad, who was basically a very kind, compassionate and generous individual. Mother would probe for the source of Caroline's outburst, reason with her, and attempt to sympathize with her and understand her problem. But ultimately, frustration, impatience and exhaustion would set in, requiring another round of professional psychiatric treatment.

These traumatic periods affected my family greatly and often caused destabilizing bouts of conflict between Caroline and me. She was hospitalized twice during the year, once in the private St. Vincent Hospital, the second time at the state hospital in Las Vegas, New Mexico. This last internment was particularly unsettling because of the tragic mental condition of the residents. I was the one who had effected this voluntary commitment. It was trying on Caroline and on me. Later in the year, Caroline calmed and moved to her own apartment. This gave her a sense of independence and dignity. A peaceful period ensued.

Rose Cruz, our trusted maid, was for Jeanne a capable housekeeper, nanny, friend and churchgoing companion. At my parent's home, Rose's mother, Incarnación Rodriguez, whom we called Chonita, was the long-suffering and devoted cook. Chonita's husband Patricio, a retired farmer, kept a small herd of goats on their land by their home on upper Cerro Gordo road. Pat was a red head, a *colorado*. He was stubborn and independent, and he resented the fact that Chonita spent long hours making my parents comfortable while he was alone herding his goats. I was Pat's friend, as I had herded goats and could sympathize with his occupation. I was also close with Chonita, since I had grown up under her care, and Rose Cruz was our trusted employee. Because of all these longstanding, close ties, the Rodriguez family knew the Kelly family would respond instantly if something was needed.

One weekend evening in February, Rose telephoned me to tell me Patricio was missing. She and her family were worried and did not know what to do. I drove up to their house on Rodriquez Hill and found the family, including several related males, sitting balefully in their living room. I began questioning them about Patricio's last known whereabouts, the location of the goats, and where searchers had looked. The responses led me to conclude he had to be somewhere on the property. It was dark, so we gathered up flashlights and went out to look for him. I organized the men into a line extending along the south border of the property. I took the center position and instructed the men to walk abreast in a northerly direction to the top of the hill, which marked the boundary of the Rodriguez property. My area of search followed a slight arroyo, or ditch, running south to north.

The line of searchers headed up the hill. About halfway up, I noticed something in the middle of the drainage I was following. I thought at first that it was a post, but upon arriving at the spot, I realized the "post" was Pat, dead—standing on his head. He must have slipped as he was trying to jump the ditch and at the same time had a heart attack. Or perhaps he was looking for a missing goat and, in the dark, stumbled and fell, and the impact killed him. Whatever happened, he wound up planted upside-down with his weatherworn sombrero still on his head.

Pat's death came as a shock for all of us. After we brought his body down to the house, the grieving commenced. His family treated me like a hero and I appreciated being with them at that sad time. The funeral, a large and dignified affair, took place at Cristo Ray church. After Pat's passing, there was no more goat herding on Cerro Gordo, although the Rodriguez family continued to live there.

Nuclear and space research expanded dramatically in the 1960s, driven in great measure by the Cold War struggle between the United States and Russia as each country strove to gain and/or retain superiority. This had a great impact on the insurance industry. Radiation, nuclear fallout, contamination, explosives, fire—all these were perils that could erupt into disasters in the industries pursuing nuclear and space activity. Insurance coverage was in demand and insurers needed to understand the risks and be able to underwrite coverage to protect those with investments in such activities.

In response to the need, several major insurance companies organized a seminar, "Insurance in the Atomic Age." The seminar was held at Las Vegas, Nevada, and included a visit to and inspection of the Nevada nuclear test site. At the time, nuclear devices were being tested underground there and had been tested above ground until just a few years before. Howell Earnest and I attended this seminar and visited the impact areas. Lectures were delivered at the site, as well as in Las Vegas proper. It was most interesting and frightening to learn about this awesome cauldron of power. In retrospect, it was amazing we were permitted to pick up fuzzed (irradiated) nuclear soil samples from within a test hole where a detonation had recently taken place.

One of the main speakers on the health aspects of radioactivity was Dr. Randy Lovelace, of the Albuquerque clinic that bears his family name. He was an articulate, charming and brilliant medical scientist. I had occasion to visit with him at the seminar, after which he flew to Aspen, Colorado, for a few days of skiing. When taking off from Aspen to return home, the plane's pilot became disoriented and crashed into a blind canyon, killing Randy and his wife. I attended his funeral in Albuquerque on December 18.

During the course of the year, my partner Howell Earnest suffered several bouts of poor health caused by a severely troubled back. I think he was also fatigued by trying to keep up with me and my restless energy as I endeavored to grow the agency. He also felt he was financially comfortable enough to afford retirement. In any case, he surprised me on December 27 with the announcement that December 31 would be his last active day at the agency, presumably making me the head and owner of the agency. I was concerned I had done something wrong, that I was somehow at fault and had caused him to make the decision to retire. When I questioned him at length, however, his response was simple: after forty years in the business, he was tired and just wanted to quit.

1966: The New Regime

Howell Earnest's resignation and retirement was finalized on February 28, 1966, but the process of transferring ownership of the agency was not just a handshake. We had to draw up new contractual agreements with the insurance companies we represented, once they were satisfied the new owner was competent. We also had to negotiate the financial considerations involved in the purchase of Howell's 51 percent interest in the business, and to transfer banking relationships. Retaining our customers, which was critical to the continuing success of the agency, involved visits, conferences, and account reviews. It was a busy period, but it went well. Thankfully, Howell was always fair and generous in his terms. His last official day ended with a gathering of his favorite company representatives and close friends, all wishing him well; they also expressed high hopes for me.

For the balance of the winter and into early spring, I worked "eight hours a day, twice a day." During this hectic period, I was also involved in a problem at Santa Fe Prep. The board tended to question and interfere with the school's academics, which should be strictly within the headmaster's domain. Headmaster Bloodgood objected to the interference and resigned as headmaster on May 12. Fortunately, it was not far from the end of the school year, but it was still a critical time because of final examinations and graduations. Oliver Seth, the chairman of the board, asked me to head the search committee for a new headmaster.

The committee, consisting of me, Bob Turner and Carrington Wooley, met and decided to call Dana Cotton of the admissions office at Harvard College. He had been a valuable friend when Santa Fe Prep was being formed. He maintained extensive connections with academia. He suggested contacting Ashby Harper, headmaster at the Albuquerque Academy. Ashby strongly recommended we interview David Jackson, whom he had known as a teacher at the Fountain Valley School in Colorado Springs. We followed his lead and interviewed Jackson, and in June he accepted the position. David served as headmaster for ten years and brought the school into its mature state.

About this time I began searching seriously for an associate in my insurance business. My accountant Stan Brumfield and Bill Vernon of the Bank of Santa Fe had been strongly urging me to hire a professional to help run the agency, and I agreed. The normal course would have been to scour the ranks of the insurance companies that serviced agencies such as mine. These "field men" were knowledgeable about insurance contracts and coverage, but were often weak in social graces and knowledge of the potential customers making up the market base.

During my search for the right candidate, I became interested in a young man, Gary Noss. Raised in Eunice, New Mexico, Noss had majored in business at the University of New Mexico, graduating in 1956. Noss had been very popular in Sigma Ki Fraternity while he was in college. From 1956 to 1960, he worked in the personnel department at the Los Alamos Scientific Laboratory. In 1960, he joined the Southwest Insurance Agency in Albuquerque and by 1966 had become a principal in the company. He knew the insurance business. He had married a college classmate, Anne, and they had two daughters. Santa Fe's cultural make up appealed to them and they wanted to relocate here. He signed the employment agreement in July and commenced work on August 1. He and Anne became popular with the younger business and cultural community in town and over the years prospered along with the agency.

At the end of the summer, Jeanne and I needed a vacation from the business and our active family. We flew to Newport, Rhode Island, to attend an alumni gathering at my old prep school, Portsmouth Priory. It had been twenty-eight years since I had graduated. The school had grown from a student body of ninety to three hundred and fifty, and there were new buildings and a much larger faculty. I missed my former masters, who had retired, but did know some of the older monks who were still active in the nearby monastery.

We next went to Boston to visit the home office of one of our main insurance agencies, the Employers Group, on Milk Street in a historic part of the city. Between conferences and meetings, we visited Faneuil Hall and Boston Commons, and dined at excellent seafood restaurants for which the city is known.

Our junket next took us to Norfolk, Virginia, and across the famous Bay Bridge to the Eastern Shore of Virginia to visit Jeanne's mother, Betty, on her 5-Y Farm. Revisiting this fabled area was delightful. Betty was a gregarious and generous hostess—and also a mother-in-law who was pleasant to be with. She made creature comfort a high priority in her household. We relished it, especially without the distraction of children. I spent five days playing golf, fishing, sightseeing and being overfed like a pâtéd goose. Jeanne stayed on, and I returned to Santa Fe on October 9. My diary notes, "Jeanne and I felt like a couple of honeymooners during this delightful visit."

I kept busy for the next two months with insurance renewals and account servicing. In November I paid my annual visit to an important client, the King Brothers butane company in Moriarty, New Mexico. My purpose in going was to award safety medals to butane truck drivers, but the trip also gave me an opportunity to visit ex-Governor Bruce King and his gracious wife, Alice. Moriarty was a far cry from the active social and political pace the Kings maintained in Santa Fe when in he was in office, but it was pleasant to see them in their natural, rural environment.

Bruce and Alice were seated in the office of the butane company when I arrived. Bruce had on jeans and, as usual, cowboy boots spotted with cow manure. When I entered the office, after friendly greetings Bruce asked me, "Bud, what's going on in Santa Fe?" Moriarty was quiet and he missed the action of the capital city.

The day after Christmas we headed for Guaymas, Mexico, for a vacation with our children at San Carlos. As always, the beach was a delightful haven for all of us. While there, I wrote my year-end diary entry, summing up the year: "Mostly taking over ownership of the agency, hiring Gary Noss, growing the business, automating the accounting and dictating equipment within the agency, buying two new cars, spending six thousand dollars on home improvements—a busy and, generally, a productive year."

1967: Opera House Fire

By early 1967 the cards had changed. I now was the senior member of the Howell Earnest Agency. Gary Noss was the aspiring young producer, hoping to prove his metal and gain a position of ownership in the firm. Howell would still make an appearance to check on our progress. He also was a source of information on how best to retain some of his favorite clients, who might be restive about continuing with this new management.

Fortunately, most insurance companies we represented were anxious to retain us as their agents as long as they received well-managed commercial accounts from us. They offered incentives based on how much new business we secured and on satisfactory ratios of losses to premiums.

In April the American Employers Co. invited me to an award presentation in Jamaica. Apparently, I was one of their star producers. I accepted the invitation to this stag affair, after clearing it with Jeanne. The celebration was held at the Jamaica Hilton at Ocho Rios in Montego Bay, a striking location with a pink-white sand beach rimming a beautiful, low-surf finger of the bay. The classic Runaway Bay Golf Club was available to us, as were rum, gin, coconuts, and a steel drum band that constantly furnished a loud metallic, rhythmic, foot-stomping type of West Indies music. The opulence of the resort was in striking contrast to the adjacent shantytown of primitive, thatched huts. A twelve-foot, heavy wire fence kept chickens, pigs and naked kids away from the resort, effectively separating two vastly different cultures. I was embarrassed by this disparity—but I did manage to overcome my reaction enough to participate in four days of golf, romping on the beach, and imbibing too much rum, all the while enduring proselytizing by company management.

When I returned to Santa Fe, reality soon set in with the daily routine of business and family life reasserting its prerogatives. The opera season was on and we attended *Cardillac* on Wednesday, July 26. It so happened that our daughter, Susan, also attended the opera that day, on a field trip from the Brush Ranch summer camp on the Pecos River, where she was enrolled for a summer course. That night, after the opera was over, a devastating fire consumed the opera house. Since my agency insured the opera, I arrived early the next morning to view a heap of charred ruins. The only relic left was a wrought iron banister leading up to the balcony.

Amazingly, the opera company did not miss a show. The opera management hastily transformed Sweeney gymnasium from a basketball court to an opera house, renting costumes and air freighting them from the Brooks Costume Co. in New York. The season finished out at Sweeney. I handed John Crosby a check for $132,000 in insurance money to pay for the replacement costumes rented from New York. Plans were drawn, and a replacement opera facility was soon in the making.

Meanwhile, changes were happening at the insurance agency, which was growing and benefitting from the addition of new personnel. We needed more space in a more prominent location. Fortuitously, Robert E. McKee Construction Co., the main contractor at Los Alamos, had just remodeled the historic Padre Gallegos building on Washington Avenue. McKee, who was looking for a tenant for the new space, twisted our arms. We recognized the space could be adapted to our needs and, since the pricing was favorable, we struck a deal. We moved our offices from the second floor of the Sena Plaza, where they'd been for over forty years, to the Gallegos building on August 26. The move transformed the agency.

Having a capable assistant allowed me to delegate some of the workload and responsibility for managing the agency. This permitted me to spend more time with Jeanne and the children. Susan was a blossoming teenager. The boys, both eleven, were becoming good skiers and reasonably good marksmen. Furthermore, they worked with me— although sometimes begrudgingly—cultivating the vegetable garden and caring for the lawn and flowerbeds surrounding the house. They competed fiercely with their grandfather in domino games and shuffleboard matches. Pamela, six years old, was a soothing influence amid the bedlam that usually reigned in our household.

In July our good friends Jerry and Florence Monks persuaded us to spend two weeks at the Elk Horn Ranch just west of Yellowstone Park. The ranch, operated by Barbara and Ronny Hymis, had been a favorite vacation spot for the Monks for several years. Jerry and Florence had six children. Their girls matched ours in age and were good friends. I wasn't too concerned about keeping an eye on the boys, as I felt the girls could handle them and, if not, I could.

The Elk Horn Ranch was located in a small valley beside the large and boisterous Gallatin River, near the route of Lewis and Clark, whose awesome expedition had always fascinated me. Horseback riding was one of our main activities at the ranch, but we also hiked in Yellowstone Park and had great adventures floating the majestic Madison and Yellowstone Rivers. On one excursion, as the boys and I hiked deep into the park, we encountered two huge bull moose; we were careful to detour around these commanding monarchs. Our time at the Elk Horn was an exciting vacation that inspired me to retrace the route of the Lewis and Clark expedition years later.

In September, while we were guests on Malcolm Stewart's Brazos River Ranch in the area known as the Brazos Meadows, the boys became addicted to trout fishing. The meandering stream was alive with trout. It was the perfect place for the boys to learn to fish, as there was a minimum of streamside brush to snag their lures, and the fish were more than cooperative.

After a brief lesson on fishing, I turned the boys loose, alone on the stream. Thomas went upstream, Robert downstream. Some time later, Robert returned with delight, showing his first catch. Later, Thomas arrived with a trout, but it looked odd. It was not sparkly fresh. It turned out that Thomas, having no luck in his fishing adventure, found a dead fish and decided to present it to us as his first catch. Nevertheless, both boys learned the thrill of successful trout fishing.

In October I tested the boys' endurance by hiking to the top of Glorieta Baldy. It was a tough climb, but we were rewarded by the incomparable view of the world from the top of the fire lookout tower.

In November we returned to the Sea of Cortez for another week at the Kiva cabana at Guaymas. Susan's friend Maudie Crawford joined us for the trip, but at the border crossing, we had nothing official that would validate her U.S. citizenship. It took a phone call to her mother in Santa Fe, plus a generous bribe to the custom guard, to permit her to cross with us. On our way back into the United States a week later, another border incident did not endear me to my son Robert, who had spent most of his modest horde of money buying illegal, potent firecrackers in Mexico. As we were going through customs, the guard asked me if there were any fireworks in the car. Ever the conscience of the good, I replied, "Yes," upon which the agent promptly confiscated the fireworks. Robert did not speak to me on the rest of the journey home. But just two weeks later he entered his first ski race at Red River and did very well. I lavished him with praise and was returned to his good graces. (Later, while in college, Robert, an expert skier, became the captain of the Harvard college ski team.)

On Christmas Day, before any presents were opened, Robert and Thomas served as altar boys at Mass at St. Francis Cathedral. In recognition of their sanctity, Santa Claus rewarded them with a bounty of Christmas gifts.

1968: Dramatic Times

As we rolled into 1968, international conflicts and domestic discontent characterized the times. North Korea commandeered the USS *Pueblo* and its crew. This crisis festered, along with the mounting conflict between the United States and Vietnam. President Johnson's frustration with this situation culminated in his announcement on March 31 that he would not be a candidate for reelection.

On April 4 a deranged racist shot down civil rights leader Martin Luther King. This provoked rage among the nation's black people and sparked fear of racial rioting and civil unrest nationwide. This steaming cauldron was further stirred and heated by the widespread rebelliousness of the hippy movement. To add to the trauma and discontent, Robert Kennedy was fatally shot by a paradise-seeking Muslim on June 5.

These tumultuous events, and the lukewarm state of the economy, led to political jousting that resulted in a Republican victory in the November presidential elections. Richard Nixon became the presidential heir, saddled with the responsibility of unraveling this tangled mess of knotty problems.

On March 12 I was made a director of the First National Bank of Santa Fe, replacing longtime director Bernard Spitz. At the time, it was considered an honor and a serious responsibility to be a director of a bank. In those days, the bank management initiated decisions and the actions, while it was the directors' place to advance the image of the bank in the community, to bring in new depositors, and to fulfill the statutory requirements of banking regulations. From a businessman's point of view, a directorship enabled one to examine economic activity within the area of the bank's operations. I continued as a director for over thirty years, striving to serve as an asset to the bank and the community.

My passion for planting and tending our vegetable garden had become something of an addiction, which led me to draft my children as helpers, often to the point of parental harassment. For me, gardening satisfied my underlying fondness for farming, discipline, and production. For the children, gardening meant geometrically exact seed planting, constant weeding, watering—a banishment from pursuing pleasures of their own choosing.

The climax of the gardening season came with the ripening of the corn. As the tassels developed, we used an eyedropper to place mineral oil drops in them. We periodically stripped the shucks to expose the kernels and evaluate their ripeness. Luck of the draw determined which of the children got to examine and report on the progress of the ripening corn. When the corn was ripe, the exciting news spread quickly.

At the height of the growing season, I had to turn over care of the garden to the backup team, since my whole family and I planned to attend my twenty-fifth college reunion on June 7. At Harvard, as at most colleges, the twenty-fifth year class hosts the key reunion. At this momentous occasion, the alma mater hopes to receive large and sentimental gifts from its graduates. Classmates are anxious to see their old friends and to show off their wives and children. More covertly, it's a time to evaluate the success, or the lack of it, of classmates, and to assess where you stand in your class and in your progress through life.

I had been to previous Harvard reunions, but my children had not. Susan, fourteen years old, and the twin boys, twelve, eagerly anticipated the big event. All were beginning to think about the opposite sex, college and the pro baseball games they hoped to see.

We flew to Boston and spent our first night at the Copley Plaza Hotel. The next day we moved to college dorms in Cambridge. Here we registered, picked up our class hats and began to reconnect with friends as well as classmates we never really knew. One of my college roommates, Morris Gray, chaired the reunion committee. He wore the class hat, an Australian broad-brimmed slouch hat, with authority and dignity. The choice of hats turned out to be prophetic, as the festivities took place in the wettest summer since George Washington took command of the Continental Army in 1776. It rained every day. Everything was wet. The tents leaked. The cold salmon snacks were clammy. My boys were disgusted—all big league baseball games were canceled. Needless to say, we were not saddened when we departed for the Eastern Shore of Virginia to visit Jeanne's mother on her farm.

The farm was situated on a neck of land jutting into Chesapeake Bay. To the east, a few miles across the Delmarva Peninsula lay the Atlantic Ocean, speckled with small uninhabited islands, which were the breeding grounds for clouds of aquatic birds. One day our family invaded Cedar Island. We walked along the shoreline, disturbing the nesting birds, which would reluctantly fly from their eggs to circle and dive bomb us. After a while, their attacks became so concentrated that we had to retreat off the island onto our boat, leaving the island to its horde of brooding, screeching gulls.

After several days of fishing and exploring the secrets of the shoreline by canoe—and indulging ourselves with wonderful meals and lots of rest—we departed for Washington, D.C., for an experience that proved to be a real education for all of us, especially the children. We visited the capitol, the various memorials, the Smithsonian Institution and, through the courtesy of Lady Bird Johnson, the White House. Lady Bird's brother, Tony Taylor, lived in Santa Fe and operated the Old Mexico Shop. He had arranged our visit to the White House, which in many ways was the highlight of our stay. From Washington, I departed for Santa Fe while the family continued on to New Jersey to visit Jeanne's college roommate.

There were a number of noteworthy events and anniversaries in 1968. My ambitious new associate at the insurance agency, Gary Noss, was anxious to become part owner. I was somewhat reluctant to accommodate him, but recalling my own experience with Howell Earnest, I determined that ownership would encourage Gary's interest in making the agency successful. I arranged a 30 percent ownership for him, an incentive that, over time, led to substantial growth for the business.

An important insurance account I acquired for the agency was Eberline Instrument Company. My longtime friend John Dendahl became a senior officer of Eberline in 1968. (He later became president.) The company manufactured Geiger counters and other uranium detection instruments, and it was a consultant for the Los Alamos Scientific Laboratory and the uranium mining interests in the area.

The year also saw the coming of age of the Santa Fe Prep School and St. John's College. The prep school graduated its second college-bound class, while St. John's awarded diplomas to its first crop of four-year graduates.

Of great local and immediate interest to the Kelly family was the honor bestowed on our next-door neighbor Christine Barker who was named New Mexico State Junior Miss. Later in the year, she won fourth place in the national Miss America contest. As an actress later on, Christine was one of the longtime stars in the preeminently favorite play "Chorus Line."

The Zuni Shalako ceremony was early that year, taking place November 23-24. We attended. No tragic falls occurred. We stayed up all night and, at dawn, returned to Gallup and fell exhausted into bed.

1969: The End of the Sixties

As we woke on New Year's day in 1969, we had no idea that this last year of the sixties would be traumatic—and the prelude to the equally traumatic seventies. The insurance agency, now known as the Kelly and Noss Agency, was firmly ensconced in its new, modern offices in the attractively remodeled Padre Gallegos building. My new partner, Gary Noss, something of an elitist, was building a clientele that reflected his rebellion against postwar conventionalism and his search for a more casual and trendy way of life. My clients, in contrast, were war vintage—conventional, and mostly well established in business and accepting of the lifestyle of the time.

In March Jeanne and I escaped from the winter/spring confusion of Santa Fe weather to spend a week at the Furnace Creek Inn in Death Valley. The inn was part of the Fred Harvey chain of resorts and a favorite hideaway for our friend Stewart Harvey. During our visit, Stewart told me about the impending end of Fred Harvey's ownership of the La Fonda Hotel. After many profitable years, the Harveys relinquished ownership in September, when Sam Ballen took over and formed the Corporación de la Fonda. My former college roommate Joe Downer introduced me to Sam, and for a period I served on the La Fonda board. Sam became a prominent and successful citizen of Santa Fe, as chronicled in his colorful autobiography, *Without Reservations*.

Shortly after our return to Santa Fe, we were confronted with a triple family tragedy. It was school break time, and the college-age children of our friends the Wooleys, Greers and Kenneys had been in Mexico on vacation. While returning to Santa Fe, driving at night, their car ran head-on into the abutment of one of the narrow bridges located somewhat south of Guaymas, Mexico. John Greer was killed and the Kenney girl was injured. The unfortunate driver was Jack Wooley.

My agency insured all three families. In Mexico, the law is Napoleonic, so that one who is accused is considered guilty until proven innocent. The authorities placed Jack Wooley in jail pending the Mexican investigation of the accident. With a son imprisoned in a primitive Mexican jail, and concerned about liability, Jack's father, Carrington, looked to me for relief. The Mexican auto insurance company, Seguros la Comercial, carried the insurance on the car. I was their United States agent. The key player for processing insurance claims in Mexico is the adjuster, and we were fortunate to have a competent and savvy adjuster, Gomez Felix, who was able to spring young Wooley from the local jail. Since no Mexican property was damaged, the impact of the accident and tragedy fell entirely upon the three American families concerned. Although the death of John Greer was a signal tragedy, the families involved eventually accepted the event as something that could not be reversed and went on with their lives. Later, on a trip to Guaymas, I had an opportunity to personally thank Felix for his adept handling of the claim.

Another, less dramatic change took place in September. After eight years of guiding the school as the founding trustees of Santa Fe Prep—Carrington Wooley, Jim Adler, Steve Watkins and I—retired from the board and were replaced with new blood.

My mentor and founder of the insurance agency, Howell Earnest, only 63, died from pneumonia on November 11, in St. Vincent Hospital.

As the year came to an end, new challenges and interests were brewing for me. Three of my children were entering or were in their early teens. Youth was in revolt and these young ones were participants. They didn't appreciate my semi-military form of discipline. Youth was seeking youth, and I was the one who had to change.

Furthermore, I realized that in the impending nineteen-seventies, all of my children would be in or graduated from college. This pretty well defined the financial responsibilities that lay ahead.

1970-1982

PROSPERITY AND CHALLENGE

1970: The Caldera

A caldera is defined as a crater formed by the collapse of the central part of a volcano after its eruption. This eruption phase of the process seemed to describe well the events that hit the Kelly family in 1970.

Jeanne and I had been married for seventeen years. We had three teenage children and our still sweet, problem-less daughter, Pamela, who was ten. I had been running a high-pressured insurance agency for sixteen years. Life's momentum had increased, and internal and external forces had built to a peak with no operant pressure relief valve.

I was in large measure responsible for much of the unnecessary stress in the family. Partially because of my military experience and my high expectations, I was a strict disciplinarian as a father. Impertinence, sloppiness, and begrudging responses from the children triggered my short fuse. As the children developed, if any of these behaviors surfaced, I immediately reacted to squash them. This tendency, coupled with the normal frustrations of business and, undoubtedly, the effects of frequent pre-dinner cocktails, often resulted in chaos at crucial family times.

Punishments such as scolding, banishment from the table, occasional spankings, and extra chores, were all part of the discipline I applied. Naturally, these attempts at discipline often triggered reactions from the miscreant such as self-justification, claiming innocence, or simply belligerence or orneriness.

Thomas was okay, but he had a defiant streak. Susan's response to my style of parenting took a more benign course. She liked boys and if I did not like her selection (which was often the case), we would battle. Robert was simply a boy discovering his limits by pulling pranks, in subtle and unsuspecting ways.

On the other side of the coin, camping, hiking, skiing, tennis, swimming, gardening, hunting, and fishing were activities we enjoyed as a family. So, life was a combination of conflict and harmony.

Robert and Thomas were not identical twins. They were only similar in their competitive drive and quest to outdo one another. They were caged together in the same playpen, bedroom, and, until they finished the seventh grade, the same classroom. After that, they attended separate schools, with Thomas at Santa Fe Prep and Robert in the public schools. As they approached high school, the need for college preparation became important. Thomas seemed well suited to Santa Fe Prep, and I knew it was a good school. In Robert's case, though, a boarding school experience seemed sensible. He was more pliable than Thomas, and we thought he'd adapt well.

In April Robert and I flew to Boston to visit prep schools. Upon arriving, I was unable to rent a car, as my driver's license had expired. Robert was too young to be licensed. This forced us to take trains, buses, taxis, and to rely on friends to drive us around. We visited Portsmouth Priory, my old school in Rhode Island. The reunion was pleasant for me, but Robert's reaction was not favorable. We then took a bus to Connecticut and looked at Westminster and Choate. Andover was on the list, but time and transportation limitations prevented us from visiting there. We returned to Boston, staying at the Harvard Club. Unfortunately, the bedding at the club must have dated from revolutionary days, as the mattresses were lumpy and felt like they were stuffed with corncobs. We had been lugging our luggage around from buses, cars, and trains, so sleeping comfortably was essential. The last night we were in Boston, we went to the movie *Mash*. I laughed so hard that I threw my back out. Robert managed to get me back to our accommodations, but the corncob bed was so unforgiving that the next morning I was crippled with back pain.

When we got back to Santa Fe, I experienced the complete range of medical treatments from every type of practitioner, from conventional physicians to veterinarians. I tried shots, needles, hot tubs, salves, exercise, and prayer. I slept on the floor. I tried everything to correct the problem, but I was practically a cripple for over two months. Finally, probably due to the combined effects of all the various treatments, my back recovered. So much for *Mash,* the hilarious movie.

On September 28 while we were preparing to visit Monument Valley with my parents, Jeanne received the sad news of her mother's death. Betty was found by her faithful maid, Irene, apparently dead from a sudden heart attack. Since our last visit, Betty had moved from her Shangri-La, 5-Y Farm, to a smaller property near Onancock, Virginia. She had been busy renovating the new home when she met her end.

Jeanne and I left immediately to join Jeanne's sister Betty Lou in Onancock. Funeral services were held in Virginia, followed by interment in the family plot, up north in White Plains, New York. We returned to Virginia, where the two sisters handled the necessary estate affairs. It was a sad ending to a wonderful part of our lives. The Virginia connection was memorable, but now it was all over. Jeanne was now a complete captive of the Kellys and New Mexico.

In September I joined my friend Pat Dunnigan, the owner of the Baca Location, now known as the Valles Caldera National Preserve, to witness the eruption of his first steam well, which had been drilled as part of a geothermal energy project. I had been following the project's progress for some time and it was exciting when the drill bit suddenly punctured through to a chamber containing boiling water and fiery rocks. The release of pressure caused an eruption of steam, mud and hot rocks that spewed to the surface and shot skyward. Further testing proved conclusively that there was sufficient heat, but unfortunately, the lack of sufficient quantities of underground water precluded the development of an economically feasible thermal well. It was an exciting experience nonetheless to be Pat's guest when the well erupted.

Bruce King, always a folksy and well-liked political leader, was reelected in November as the Democratic Governor of New Mexico. Shortly thereafter the year ended for me with a strong resolution: I would try to be aware that my gifted children deserved recognition, that they were becoming adults and no longer were to be molded into my image and likeness.

1971: A Year of Transition

Bruce King's inauguration in 1971 marked the beginning of another busy year, with much change. Events took place that would set the direction and pace of activities for some time ahead. This was especially true for my children, as their initial steps towards independence and self-reliance commenced.

Thomas, who had been at Santa Fe Prep for two years, longed to be out from under his parents. Also, he was an able basketball player and the limited sports athletics program at Prep offered little challenge. Friends of Thomas were attending the Abbey School at Canon City, Colorado and he persuaded Jeanne and me it would be good for him to go there, too. When we visited the school, Thomas was enthusiastic and made it clear he had no interest in investigating eastern boarding schools. I felt this Benedictine school was a poor educational choice, especially compared to Portsmouth Abby, my school. But despite my misgivings, he applied to the Abbey School and was admitted for attendance starting in the fall.

Robert was not as restless as Thomas, and he was also more studious. He had an obvious artistic talent and had already begun painting. Although I had exposed him to eastern boarding schools, he had shown no great desire to move to the East. Also, he was an accomplished skier and raced for the local ski team. Adding all this up, and with our concurrence, he elected to stay at home and enter Santa Fe High in the fall.

Susan was to be a senior at Santa Fe Prep. She had attended the school for four years and was happy and doing well there. Her sights were set on college admission, so no change was indicated. Pamela, who had only just entered grammar school at Acequia Madre Elementary School, presented no problem. Her time of maturation and restlessness was yet to come.

While the children were on summer vacation, keeping them productively occupied was a necessary self-defense for us, their parents. When I was their age, I was set loose with a horse and an older brother to spend the summers like vagabonds in the mountains or on ranches. Jeanne, raised in New York, spent her summers at traditional camps. Our children had a mix of both experiences, but this year we decided each of our older children would spend several weeks away at a camp involved in character building activities.

Susan was mostly interested in attracting boys. She needed to be challenged physically and spiritually. We enrolled her in the Colorado Outward Bound school in the San Juan Mountains for five weeks of hiking, camping, rock climbing and canoeing, all the while learning how to work with a team to succeed. The final test was surviving for several days, totally isolated on a "solo," subsisting on what could be gleaned from nature. At the end of the course, I flew to Denver to meet her. When she emerged from the bush, I was delighted to be greeted by a robust and self-confident young lady.

The boys required tempering also. Being competitive twins, they needed to be separated. Each was interested in different activities. Thomas was a competitive athlete. He liked tennis, and I enjoyed playing with or against him. He chose to go to John Newcomb's tennis camp at New Braunfels, Texas. There, he spent three weeks honing his game and learning to be a good sport. On his return we played frequently; he relished beating his father.

Robert was interested in scuba diving and viewing aquatic life. Dry New Mexico was not the place for this, so he spent several weeks at a sea camp in Florida. It was a good choice for him, particularly because it forced him to be on his own, away from his twin and free from parental direction.

In June, Jeanne and Pamela enjoyed a change of scenery when they attended the wedding in New Jersey of Carrie Schultz, daughter of Jeanne's college roommate, who was marrying Laughlin Barker Jr., son of our next-door neighbor in Santa Fe. The two had fallen in love when Carrie was our houseguest the previous summer. Jeanne and Pamela both claimed credit for this romance and were honored as cupids at the gala wedding ceremony.

On May 2 the annual Amigos trip commenced. Governor King was the honorary leader of the crusaders who were venturing forth to spread the charms of New Mexico. Bill Vernon, Phil Register, Leland Thompson and I were among the Santa Fe contingent. Our lengthy trip began with a stop in Oklahoma City, where we toured the Kerr-McGee oil well servicing and refinery facility. From there, we flew to Austin, Texas, intending to visit with ex-President Lyndon Johnson at his Pedernales Ranch. When our bus-driven contingent reached the ranch gate, however, security would not let us enter. Our visit had not been cleared by the FBI. As a result of this snafu, LBJ was no longer one of our favorite politicians.

The next leg of our trip led to Memphis, Tennessee, and then Charleston, South Carolina, where we met General Mark Clark, who was then the head of the Citadel Military College. I had served under him when he was the Commander of U.S. Forces in Austria in 1945 and 1946.

After the stop in Memphis, the Amigos flew to Freeport in the Bahamas Islands, as a recreational dividend. Here we spent two days enjoying the tropical island and its famous beaches and coral reefs. We covered five thousand miles on the trip, returning to New Mexico on May 9.

Two severe insurance claims had to be settled during the summer: one was the hail damage to the Dixon apple orchard. The other was the fire loss of the Honnell store and home in Tesuque.

Two family hikes were notable. In August we hiked from the ski run across the mountains via the Winsor Trail to Cowles on the Pecos River. I had previously done the trek on horseback and skis. This was the first time on foot and with the entire family.

Another adventure was climbing the sacred Black Mesa near San Ildefonso Pueblo. The mesa top supposedly was used by the pueblo for puberty rites. In September, as a birthday present for Susan, we climbed to the mesa top. Susan had as guests a mixed group of her young friends. Once on top, we had a picnic and a pretend initiation into adulthood. We were fortunate to be able to visit the mesa, as later the pueblo prohibited access to the mesa top.

My junior partner, Gary Noss, was honored when he was sworn in at Albuquerque as Honorary Consul of West Germany on November 9. He was appointed by Dr. Richard Lonis, consul general, who was stationed in Houston, Texas. Prophetically, my diary noted, "This is a big day for Gary. He loves anything that smacks of status and élan. I am basically more of a fundamentalist. As long as it doesn't interfere with his job and the agency – okay!"

On Christmas, we celebrated with a family dinner at my parents' house. Following it, Jeanne and I and the children headed for Taos Ski Valley to spend the few days remaining in the year. Here we enjoyed a festive and snowy New Year's Eve to end this transitional year.

1972: A Business Goal Achieved

As I began my eighteenth year in the insurance business, the agency was strategically situated in the historic Padre Gallegos Building on Washington Avenue, two blocks from the Plaza, a prestigious location in the center of Santa Fe. Known by this time as the Kelly and Noss Agency, the firm represented the best insurance companies in the United States. Our customer clientele, both individual and commercial, included an enviably successful segment of the area's economy.

Our successes didn't come about magically. To survive and prosper required hard work. Customers were demanding, competition was relentless, and professional competence had to be progressively maintained in the office staff. The work was constant; like medical professionals, we had to be available to service on claims and accidents at any time of the day or night.

In addition to the demands of business, I carried responsibilities with other organizations in Santa Fe. The combined stress from all these activities was taking a toll. In March I realized I was really tired. I developed a hernia, which Dr. Smithpeter cinched up at St. Vincent Hospital. This took me out of circulation for about two weeks. (When Smithpeter was done, he said, "This stitching job will never come loose." It never has.)

Around that time, I received some news that accelerated my convalescence: the American Employers Insurance Co. invited Jeanne and me to a company gathering in Portugal. On April 13 we flew to Lisbon and for a week we were toured, fêted and royally entertained as we visited Lisbon, Estoril, Fatima, Nazaré, and Sagres. After a final dinner and dance at our hotel in Estoril, the tour officially ended and we were on our own. We decided to extend the trip by visiting our friends the Dillons in Sevilla, Spain.

We drove south to the southernmost region of Portugal, the Algarve, through beautiful, green, well-tended grain fields and pine and eucalyptus timber stands to Lagos, 175 miles south of Lisbon. We toured this picturesque coastal area for two days. Portugal in 1972 was more prosperous than Spain, and the longstanding relationship between Portugal and Great Britain had cultivated a cordial openness towards foreigners. We found Cabo de São Vicente, the most southwesterly point in Europe, to be especially impressive. The stunning view of the South Atlantic Ocean from the promontory prompted us to think about Prince Henry the Navigator and other Portuguese maritime explorers and their historic voyages of discovery.

From the Algarve we toured eastward, crossing the Guadiana River into Spain and on to Sevilla. The Dillons had been in Spain for several months and served as pleasant guides for the next few days. We then drove northwest back into Portugal and to Lisbon, where we departed for the United States. It was a delightful, restful trip, and primed us for the return to home, family, and work.

Susan graduated from Santa Fe Prep in June, Thomas returned from boarding school, and Robert finished his first year at Santa Fe High. It was job time for them. The boys found construction jobs with Mr. Marr, who was building Road Runner Homes, the first condominium complex on the Old Taos Highway. I was delighted they were learning the construction trade from a demanding but friendly veteran of the building industry. Susan was employed at David Agnew's Aspen Country Store.

In September Susan and I flew to Boston to enroll her in Pine Manor Junior College, her mother's alma mater. Our fellow passenger en route was John Pen La Farge, on his way to Boston University. My father was Pen's godfather, and he took the responsibility seriously. He was disappointed when Pen did not choose Harvard. While Susan was involved with admissions paperwork, I visited the companies my insurance agency represented. I also saw Fred Jewett, the dean of admissions at Harvard, whom I was planning to help on his annual fall recruiting trip to New Mexico.

As the business year was nearing a close, all of us at Kelly and Noss were striving to reach the a goal of one million dollars in premium income for the company. This kept us very busy. I did, however, manage to fit in two family adventures. We spent a long weekend at Chinle, Arizona, exploring Canyon del Muerto and Canyon de Chelly. On the visit I remembered a photograph my parents had taken of D.H. Lawrence and his partner Frieda Weekly hiking in the canyon years before. The other adventure was climbing Pedernal, Georgia O'Keeffe's sacred mountain. I knew Georgia and thought this achievement would impress her. (She didn't impress very easily.)

It was snowing hard when Fred Jewett arrived from Harvard. I drove with him through the snowstorm to Los Alamos High School, where he spoke to students interested in attending Harvard. I think they were as impressed by his adventurous spirit as much as they were by the college he represented. From Los Alamos, we visited Santa Fe Prep, St. Michaels, and the Santa Fe High. As I recall, several students from these schools were admitted for the next school year.

The business year ended successfully, as we surpassed our one-million-dollar goal.

1973: Trauma and Release

In January 1973 the ski season was in full swing when Rufus Barringer, a ski area insurance mogul and my primary contact in the industry, arrived from New York City for his annual underwriting inspections. We skied at each of the ski areas on his schedule—Taos Ski Valley, Lake Peak (Santa Fe), and Los Alamos—inspecting chair lift equipment, trails, and signs, and reviewing safety precautions. We advised management about new developments in equipment, legislation and legal liability claims and judgments. We also entertained the ski area operators with dinners and social gatherings.

On the national stage, President Nixon was serving his second term. He optimistically announced a Vietnam ceasefire and an agreement that would end U.S. participation in that unpopular war.

School spring break in March came just when Jeanne's college roommate Mickey Schultz and her husband Jack invited us to spend time on John's Island, near Vero Beach, Florida. They had just moved to this newly developed beach resort. In addition to spending time with the Schultz family, we visited with Bob and Natalie Dodge and their three sons in Vero Beach. Bob, the pilot I had worked with in uranium prospecting, was an avid sailor. His pet was his small yacht, *Nattie*, which he used to navigate the local rivers and the ocean. It was a novel experience for my family to be on board.

From the ski slopes of Santa Fe to the Florida tropics, sports diversity was the drill. Swimming, beach activities, golf, tennis, and boat cruising filled the days. Generally, the weather was ideal, although a storm system came in with heavy rain and wind on March 17, forcing a halt of outdoor activities. My parents had just finished a cruise in the Caribbean and were staying at the Sea Ranch in Fort Lauderdale and visiting the Hootons, who were living there. I called my father. Since the bad weather limited our activity, we decided to drive to Fort Lauderdale the next day for a parental visit and a late Sunday lunch.

Running a little late, I called to let my parents know we would arrive in about a half hour. Barbara Hooton took my call. Hesitatingly, she said, "Your father is dead. He died of a heart attack while swimming off the beach." Upon arriving, I found my father was indeed dead—but what a classic death. He and his friend, Dr. Lonsway, had attended church that morning and then, after a fine breakfast, decided to spend a little time lolling on the beach. Dr. Lonsway noticed that my father, who was floating whale-like in shallow water, was drifting a bit too far from the shore. Dr. Lonsway called out but received no response. He swam out and pulled my dead father onto the beach.

By the time we arrived, the legal investigation of the cause of death had been taken care of and my father's body lay at the mortician's. I initiated the next steps in dealing with his death, arranging to ship his body to Santa Fe and assisting Mother in preparing for the flight home. Jeanne and the children returned to John's Island. The following day, Mother and I and flew to New Mexico with the deceased.

My brother Booker and I made the arrangements for the funeral and interment in Santa Fe. I recall we selected one blue Brooks Brothers suit and a Harvard College tie to have our father fittingly dressed, as he would have chosen for his last journey. The funeral Mass was celebrated at St. Francis Cathedral on March 22, with interment following in the family plot at

Rosario Cemetery. As the flag-draped coffin was lowered, the final military taps brought a lump to my throat and to all of the large group of mourners. At home that evening, the family and the veteran household help had a final dinner for Dad, which included a toast to me on this, my fifty-second birthday.

For the next three months, my mother gradually acclimatized to life without her husband, and we adjusted to being with her in a new way. All the while, trauma with my sister Caroline was mounting—a long and stressful buildup that lead to her tragic death.

At the time of my father's death, he and Mother still lived in the family home at 531 Palace Avenue. My unmarried sister, Caroline, had the guest house adjacent to the main house. Only a shared driveway separated my house from Caroline's. Thus, Jeanne and I were in the epicenter of the family universe and became responsible both for Mother and Caroline, who was increasingly disturbed. I was the executor of my father's estate and thus was intimately involved in my mother's affairs.

My mother's finances became an immediate concern. Keeping up her big house and its grounds was expensive, and the house was much larger than my mother required. It seemed the guesthouse occupied by Caroline would be the logical place for Mother to live. This would free the main house for rental or sale, but it did involve displacing Caroline and finding a suitable rental for her elsewhere in town. When we presented Caroline with this, the most reasonable solution to Mom's housing problem, she suffered an emotional breakdown. We weren't entirely surprised, since she had had a history of temperamental highs and lows most of her life, but as the deadline for her to move out of the guest house loomed, she became increasingly depressed, resentful, and somewhat irrational.

I got a break from the situation when I joined the Amigos for the annual trip to promote New Mexico. This year we had an ambitious agenda. Our large group, which included Governor King, visited Indianapolis, Washington, D.C., colonial Williamsburg, and Atlanta, where we met Governor Jimmy Carter. I was impressed with Carter's considerable warmth and charm. From Atlanta we flew to Puerto Rico and then returned to New Mexico, exhausted.

By the middle of June, it was time to make the housing changes, but Caroline had not yet found a new home. On June 20 one of Caroline's old boyfriends, Jimmy Overall, arrived. He had been recently widowed and was anxious to rekindle his friendship with Caroline. He recognized her emotional problems and wanted to help.

While having dinner at Caroline's house on Saturday night, June 23, Jimmy suggested they take a trip together, hoping this would be a morale builder that might re-spark their relationship. He suggested they leave on Monday. After dinner, he mentioned he was going with me on the next day to my mountain retreat at Cow Creek, our small ranch adjacent to the stream carrying the same name. Caroline wanted to go too, but he felt they needed a few hours apart. He said he would like her decision about the trip when he returned Sunday afternoon.

Cow Creek was an elixir for Jimmy and me. We talked about Caroline and both felt refreshed when we returned home, where we were met by a worried Jeanne. She hadn't seen or heard from Caroline all day, although her car was parked in its usual place by her house. Concerned, Jimmy and I knocked on her door and called out. No answer. I unlocked the door and found Caroline dead in her bed.

Caroline looked peaceful, lying there in her nightgown with her clothes neatly placed on the bedside chair. On her bedside table we found a vial of sleeping pills with only two remaining in it. It was clear she had been dead for some time, as rigor mortis had set in. Jimmy fell to his knees to pray. I called Dr. Angle and Caroline's favorite priest, Fr. Shower. Dr. Angle arrived, pronounced Caroline dead, and, at my request, arranged for an autopsy. The padre arrived to bless the body, which was then taken to the mortuary.

Caroline's funeral was on Wednesday, June 28, at Fr. Shower's church. Fr. Shower was a Vatican II reformist, a broad-minded priest who had become a good friend of Caroline, and she his good parishioner. She was buried next to her father at Rosario Cemetery.

Caroline's death was a tragedy. She was gifted in many ways—bright, attractive, entertaining, adventurous, artistic, and gregarious. These attributes, however, were mostly evident outside of her immediate family. For whatever reasons, her more unattractive traits seemed to bubble up to the surface when she was bored or thrown into a competitive position with her parents or siblings. She could be critical, vindictive, and abusive towards her family. At the same time, she was remorseful, self-pitying, indecisive, and unhappy with herself. Despite the best efforts of my parents and considerable attention at medical facilities, her extreme highs and lows persisted most of her life. However, even though she was obviously troubled, I don't think she took her own life.

I believe her depression over Dad's death, her fear of moving to another home, and indecision about Jimmy led her to desperately take a large dose of medication, which turned out be an overdose that killed her.

Caroline's untimely death came as a blow to all of us, but at the same time it was a release. It may seem heartless for me to say so, but she had been making life so difficult that, once she was gone, the pressure ceased and we felt a sense of relief. And remarkably, in a kind of celestial restitution, a life insurance policy Caroline had purchased years before gave Mother, as beneficiary, sufficient finances to allow her to retain her home until her death.

The balance of the year was relatively peaceful. Jeanne and I felt a sense of calm companionship. Our main concern now, other than our children, was caring for Mother, our next-door neighbor.

1974: Relative Tranquility

The ballgame of life played out in our favor in 1974. We had more hits than strikes. The children were accepted by the colleges of their choice. The agency had a profitable and relatively calm year. Jeanne and I continued our busy, parallel sets of interests and activities.

The boys turned eighteen in April. Thomas was set to graduate from the Abbey School in Colorado. He was on the honor roll and had been counselor at the school's summer camp. He applied to the University of Arizona and the University of San Francisco, a Jesuit school founded in 1855. After visiting both campuses, he selected San Francisco, I think because of the lure of that attractive city and its swarming hive of hippy oriented youths—not the academics. Jeanne and I accepted his choice.

Robert was finishing at Santa Fe High School and looked eastward for college, applying to Dartmouth, Bowdoin, and Harvard. Dwight Miller, the Harvard recruiter, had worked hard to recruit him, as had I. We were pleased when he was accepted at Harvard. Susan, who had graduated from Pine Manor Junior College, transferred to Colorado College in Colorado Springs. Pam was in her first year at Santa Fe Prep.

In September when the three collegians left the nest for their respective alma maters, the rapid current of activities on the home front subsided to a meandering flow. On weekends I usually went to Cow Creek, which I found to be a peaceful haven and an escape from the bustle of Santa Fe life. Thinning the pine stands and clearing logs and fallen trees was a demanding yet satisfying activity for me. Jeanne often would accompany me, along with my hardworking and knowledgeable woodsman, Gary Valencia.

Jeanne led a double existence. Our social life involved considerable business entertaining. The several boards I served on had functions that required her involvement as a guest or hostess. Our group of local friends met frequently. This combination of social events absorbed a considerable part of her busy life. Of course she was a devoted wife and mother and she also had a number of interests of her own, including seminars at St. Johns, church study groups, a real estate course, running the Opera Store, and, in season, skiing and tennis. In September, while playing tennis with Ethel Ballen, Jeanne fell and broke her left wrist. This accident put her out of athletics until late in the ski season. She substituted a tennis racquet for a St. John's seminar until the wrist healed.

The insurance agency had a profitable year. My partner Gary felt the agency needed to recruit and groom a younger man to be a replacement when one of us retired or left the agency. I agreed. My interest focused on Charles MacKay. His father, Jack, had been a key employee of Gross Kelly & Co., and I knew the family. Charles had finished college and seemed to have the desired potential. He was young, handsome, well spoken and knew the Santa Fe area. The only trouble was his great interest in music. He was an accomplished French horn player, had an extensive knowledge of classical music, and was a fan of the local opera. These were positive attributes, but they were not essential for a successful career in our business. His father agreed. Charles eventually realized this, too, and despite my considerable effort to entice him, he decided to seek a career in music. John Crosby, director of the Santa Fe Opera, hired him as an apprentice musician. His talent led him to the top of his profession. (As I write this in 2010, he is now the managing director of the Santa Fe Opera. Charles made the right decision.)

We kept looking for the right person. Gary brought up Ray Pineda, a fellow University of New Mexico alumnus, as a likely candidate. Ray was then employed by Xerox and traveled often to Brazil, but he wanted to settle in New Mexico on a permanent basis. I agreed to consider him on his next visit, which was to be in 1975.

One reason we needed another man in the business was my preoccupation with board member responsibilities with the St. John's College, the First National Bank, and the School of American Research—all of them clients of the agency. In retrospect, I was guilty of the sin of participating in conflicting interests, but this didn't seem to be a serious matter then. I felt I could serve the interests of these organizations effectively, since I was familiar with their operations, and I was able to obtain for them the lowest insurance pricing available. In any case, my fellow board members apparently felt I was a desirable asset, which I strove to be.

1975: The Rift

In early 1975, as chairman of the board of trustees of the School of American Research (SAR), I found myself entangled in a tricky situation. The Museum of New Mexico, a state institution, and SAR, a private institution, were engaged in negotiations regarding the ownership of their archeological collections. The museum questioned SAR's ownership of part of the collection, and visa versa. Originally, SAR and the Museum of New Mexico shared collections, but, with the passage of time, the two institutions became separate and distinct, and ownership of the artifacts became contested. SAR felt it was essential to brief the new governor, Jerry Apodaca, about the existence of SAR and the conflict over the collections.

Dr. Doug Schwartz, the president of SAR, and I invited Governor Apodaca to a luncheon meeting and tour of the school. The governor was impressed by the striking buildings and grounds (the former Elizabeth White estate). We briefed him thoroughly on the history of the collection ownership problem.

Later, to further impress the governor, who was an avid sportsman, we invited him to the Kiva Club, where both Schwartz and I were members, to see a squash game. Our friendship developed and Apodaca was made an honorary member of the club. He was anxious to learn the game of squash. I gave him his first lesson. Since he was a tennis player, he had a natural affinity for squash. After a few rallies, he wanted to play a real game, and I obliged him. At one point during the match, he failed to give way to my shot, and the ball hit him. He fell to the floor. His competitive instincts were aroused and he wanted revenge. He came up swinging but he calmed almost immediately. Apodaca became a good player. (The negotiations between SAR and the Museum of New Mexico ultimately resulted in a settlement, although it took longer than Apodaca's single term in office before this was accomplished.)

Because of our successful business year, the Commercial Union Insurance Company invited Jeanne and me to an award week in Monaco, Monte

Carlo, of Princess Grace fame. This toy-land principality of six-hundred acres, located on the coast of France, gave us a wonderful exposure to an unreal world of casinos, beaches, yachts, palaces, changing of the guard—all of it a charming Disney-like fantasy, featuring Grace Kelly as the star of the production.

On June 1 we departed Monaco and drove via Marseilles up the Rhone Valley and on to Paris. En route we stopped at Avignon and, after crossing the *pont*, spent the night at La Prieuré, a converted monastery. The next day we visited the Palais des Papes, the palatial home of the popes during the Babylonian Captivity from 1309 to 1377. Here we saw a retrospective exhibit of Pablo Picasso's paintings, from the sublime to the grotesque.

From Avignon we detoured to Cluny, the cathedral town where the French Benedictine order once had its headquarters. This was interesting to me because I was schooled by English Benedictines. Coincidentally, at the time of our visit in Cluny, a reunion of the French Resistance fighters was taking place there. These underground guerilla heroes of France during the German occupation were recounting their exploits against the Bosch (the German army), and consuming lots of wine. I joined them and, as a former U.S. Army officer, was treated as a hero. As they say, "Old soldiers never die, they just smell that way."

After I was awarded the title of honorary member of La Resistance, we proceeded up the Loire Valley. As we passed through Clermont-Ferrand, I silently saluted the memory of Archbishop Lamy, the first bishop assigned to the New Mexico Territory after the war with Mexico. His birthplace is the location of iron mines that date back to Roman times.

When we arrived in Paris, we were greeted by Peggy Catron of Santa Fe, an exchange student. She was our guide for three adventurous days in that splendid city. Our trip ended June 6 when we returned to real life in Santa Fe.

Before immersing myself into agency affairs again, I spent some time at the Cow Creek property. Robert, on summer vacation from college, was engaged in a logging operation there. A forest fire long before had stimulated the growth of a heavy stand of aspen trees. Over time, this nursery growth had aged and had been replaced by conifers. Dead aspen trees littered the area, but the trees were sound and, when cut into eight or twelve foot lengths, were in demand as tunnel supports for the uranium mines in the Grants, New Mexico, area.

Robert and his pal Paul Hart camped on the property and harvested the aspen. They felled trees, lopped of branches, cut the trunks into proper lengths, and hauled them to a decking area several miles from their camp. Here they were loaded onto a sixteen-foot semi-trailer and trucked some two hundred miles to the mine sites.

The primitive road into and out of the decking area was a challenge for John Hite, the trucker, who bought the logs from the boys. During the course of the summer, well over the equivalent of one linear mile of logs was harvested and shipped to the mines. Robert and Paul made some hard-earned money, the mines obtained the timbers, and the Cow Creek property was cleared of the heavy downfall.

Our ski area business was threatened by serious competition. The insurance market I had used was a syndicate of insurers located in New York City; the marketing manager was a company named Barringer and Williams. This company had a virtual insurance monopoly for the ski area operators in the East, as well as the scattered ski areas in the Southwest.

As the skiing industry expanded westward, Pacific Northwest ski areas organized the Western Ski Association, a trade organization that offered an insurance program. In order to compete against Barringer and Williams of New York, the Western Ski Association's program offered insurance at a lower premium and with broader coverage. Because of my loyalty to Barringer and Williams, I resisted contracting with this rival insurer, and as a consequence I lost market share in the industry, including my longstanding and highly valued accounts with Santa Fe and Taos Ski Valley. I knew also I would lose other operators in New Mexico and Arizona unless I could match this competition. I made a monumental effort but I could not budge my market to meet this competition. A rash of liability claims against the eastern operators had made the existing pricing unprofitable. Rather than lower rates, an increase in premiums was inevitable. I knew this would ultimately finish Barringer and Williams in the West. Emotionally as well as financially, this development was a severe blow.

Another festering problem surfaced in June. Gary Noss, my partner now for several years, had an emotional crisis. He abruptly left his wife, Ann, leaving her a note saying he was depressed and had decided to leave his family. After a frantic day of pondering what to do, we received a phone call from Flagstaff. It was Gary, saying his mind had cleared and he was returning home to his wife, family and job. We were relieved, happy, and forgiving. He resumed his usual routine at work and appeared cleansed from whatever demon that had possessed him. I had lingering doubts but assumed a positive attitude.

One of the complicating factors in this scenario was the imminent arrival of Ray Pineda, Gary's protégé, as a new, young and potentially vital addition to the agency. Gary had been impressed with Ray as a student at the University of New Mexico. Ray was bilingual and, after graduating, was employed by Xerox. He had advanced with the company and had been traveling often to Brazil, but wanted to settle in New Mexico. Although somewhat dubious, I acquiesced to Gary's desires to hire this young man. Ray accepted our proposal. He was scheduled to start on August 17 as a trainee under Gary's supervision. He arrived, and after briefings and some exposure to Gary's area of responsibility, he departed for a six-week insurance course in Philadelphia conducted by the Insurance Company of North America. He rejoined the agency in late October.

On November 3 Gary announced his "declaration of independence." He entered my office and stated, pursuant to our agency agreement, he wanted to retire as of the first of the year. He stated his marriage was on the rocks and he wanted to bail out and start a fresh life.

This was a bombshell but not exactly a surprise to me. He had been acting deviously for some time, taking unannounced trips, meeting with the German Consul General in Houston, and participating in other activities that annoyed me. His actions had disrupted our communication and interfered with the proper flow of business. Obviously, his departure would be a further disruption, and it left Ray Pineda out on a limb. It also made me financially obligated to buy Gary's substantial interest in the business.

All the upheaval led to a major overhaul of the business and caused much emotional trauma to Gary's family and to me. We drew up separation agreements and initiated accounting measures. Gary divorced Ann and, finally, on December 19, he officially departed. Shortly thereafter, Gary formally announced he was gay. He changed his appearance and circle of friends and ultimately landed in San Francisco to join the gay culture for which that city is still famous. Gary's exit from conventionality was defiantly blatant. He must have become desperately discontented acting in his conventional role. It took a fierce act of courage to discard the old and convert.

Now I faced the task of severing the legal and financial bindings that had been in effect for almost nine years, reforming the agency and getting back to a normal business life. The year ended with this rift in the agency lingering wide and dark, promising many challenges in the new year.

1976: Reorganization

By the end of January 1976 the lawyers, accountants and bankers had finished their tasks, completing the formal "divorce." Gary resigned from the agency, I assumed the obligation to pay him for his portion of the agency's equity. I was free, though burdened with a substantial debt.

Another responsibility I inherited was Ray Pineda. Gary's bombshell was a great blow to me, but its effect on Ray was greater. He was left in a business venture that was new to him, without the support of his mentor. Unconsciously, I associated Ray with Gary's desertion. In retrospect, this seems unfair. I went through the motions, complying with the stipulations of Ray's employment contract. I furnished him with the best training available and, ultimately, he became a minority owner of the agency. Still, our association never gelled into a genuinely harmonious relationship.

Once Gary's separation from the agency was complete, I immediately moved to eradicate any vestiges of our association. The fallout from his precipitous and ill-timed departure had to be purged. My first move was to eliminate any connection with the German vice-consulship, a relationship Gary had initiated and maintained. I removed from the agency's entrance the elaborate German consular plaque that had been given to Gary. I then legally changed the name of the business from the Kelly and Noss Agency to simply the Kelly Agency. I assigned Ray the responsibility of servicing and monitoring most of Gary's accounts and realigned the agency's internal organization, appointing Joanne Legits as office manager. Relying on Ray's experience in automation, which he learned at Xerox, I initiated a conversion to more modern control procedures.

The turbulence of making these adjustments smoothed out in time, and we gradually resumed normal operations. Despite the Noss separation and the wrenching loss of several key accounts like the Santa Fe Opera and Taos Ski Valley, 1976 turned out to be a profitable year.

I remained very engaged with the affairs of the School of American Research as I served out my second and last year as chairman of the board. Dr. Doug Schwartz, SAR's president and CEO, was the dynamo behind the school's growing reputation in southwestern archeology. Most of the board's energy continued to focus on negotiations with the Museum of New Mexico over the legal ownership of the collections claimed by both organizations. We made much progress towards a solution, although the problem was not settled during my term.

I was privileged to become acquainted with Elizabeth White, a friend of my parents. A gifted and intelligent woman, Elizabeth was a major benefactor of SAR and had conveyed her palatial home and grounds—as well as her property in Sena Plaza, downtown—to the school. Her former home was the epicenter of the school's operation during my tenure.

One of the highlights of the year was the two-day SAR seminar at Chaco Canyon in October. Jeanne and I were the guests of Doug and Anita Schwartz, sharing their rented van for the occasion. Later that month, at the annual meeting of the school, my tenure as chairman of the board ended and Dr. Phillip Schultz took over the reins.

It also was a busy year for the family. Before the older children went back to school, we hiked to the top of Tetilla Peak. Afterwards Thomas departed for a year abroad in Italy, while Susan returned to Colorado College for her final term and Robert went back to Harvard. Pamela, the only child left in the nest, returned to Santa Fe Prep.

We went skiing on Jeanne's fifty-fourth birthday, (which, incidentally, fell on the same date as George Washington's, February 22), staying at the Tamarron Lodge in Durango, Colorado. I enjoyed the wide and gentle slopes at the Purgatory ski run, while Jeanne, more adventurous, traversed its more demanding runs with Hank and Bonnie Kelly, our nephew and niece, who acted as our hosts.

In June Robert completed his second year at Harvard and was asked to serve as a senior leader for the National Outdoor Leadership School (NOLS). He had spent the previous summer ice climbing with NOLS in the Cascades. This time he joined an expedition to climb Mount McKinley, or Denali, the highest mountain in North America, some 20,320 feet high. The mountain has earned a savage reputation for being a dangerous and tricky summit to conquer. The ascent took Robert and his crew over forty days, involving several base camps. The team endured frightful weather. Robert climbed with the final assault team and gained the summit, despite detouring on a rescue mission to help an unsuccessful party of summit seekers.

In August I took my mother and several of her friends to her favorite religious ceremony and annual picnic, the Feast of Our Lady of the Angels, held at the mission church at Pecos Pueblo. Each year, the descendants of the Pecos Pueblo people, who have lived at Jemez Pueblo since the 1830s, return to their ancestral home. The return coincides with an annual procession of local people carrying the painting of Our Lady of the Angels from the parish church in Pecos town to the Pueblo mission. This ancient Mexican painting, formerly lodged in the old Pecos Pueblo mission, is returned there for the occasion. The priest celebrates Mass and the Jemez delegation is honored. After the Mass and a fiesta, the painting is returned to its adopted home in the Pecos parish church.

After that August celebration, the rest of the year was, thankfully, quiet. Christmas brought the family home, except for Thomas, who elected to extend his time in Europe for a few more months. On Christmas day he attended Mass at Saint Peter's in Rome and we did the same at Saint Francis Cathedral in Santa Fe.

1977: Intermezzo

On New Year's Day Jeanne and I drove to Leland Thompson's La Bajada Ranch and walked down the Rito San Antonio to the Santa Fe River. It was a peaceful, almost spiritual walk. We had no great problems facing us. The children were settled in their schools. Despite all the buffeting in 1976, the agency had completed a satisfactory year. Our health was good, and we felt we could handle the challenges facing us.

I would characterize 1977 as an intermezzo, a short, light and generally pleasant *entr'acte* between more turbulent times. But despite this reflective observation now, at the time I wasn't so sanguine. It was more my nature to take a pessimistic rather than a positive, confident view of life and its challenges. Worrying was my specialty, and it was always an obstacle to overcome. In fact, my worry often presented more difficulty than the problem that caused the worry. I realize now this tendency warped my attitude and had an effect on the people around me. If someone did not share my apprehension concerning the gravity of a problem we faced, I would resent it and let my pique be known.

As the head of a fairly large insurance agency, I had my share of problems. Personnel troubles were constant. Each of my staff, which consisted mostly of women, brought to work a unique temperament. They also brought with them home pressures, personal financial concerns, and other baggage. In the work place, emotional upsets arose, people nursed jealousies and rivalries, and insecurities often came to the fore. It was something of an art to manage all these human foibles to orchestrate a harmonious and effective working organization.

Then there were the relationships with clients, which lie at the heart of the insurance business. The very nature of insurance—safeguarding assets against the risk of loss—makes people resent having to buy it, and they often regard the purveyor (the agent) as a whipping boy. Securing clients and retaining them is a twenty-four hour-per-day job—a task that kept me in the frying pan constantly. Some clients were always difficult. Key accounts, such as motels, shopping centers, banks and retail outlets, required continual massaging. I would constantly worry and fret.

On top of all this, we faced fierce competition, especially for ski industry clients, which were constantly targeted by competing insurers. I lost some accounts but fortunately retained most. My new associate Ray was plunged into this fray. Despite being a novice, his easygoing, laid back and friendly temperament was a welcomed balance to my intensity.

In the spring Jeanne and I took advantage of a calm in the business storm to take a trip to Bermuda, accompanied by Pam, who was on spring break. Our first stop was New York City. Pam and Jeanne checked into the Harvard Club after letting me out at the Santa Fe Opera office at 48 East Sixty-third Street. Kenneth Beers and Charles MacKay met me there and showed me the facility, which was John Crosby's New York home and the audition center for the Santa Fe Opera. I visited briefly with John, reviewed his insurance coverage, and then walked to the Harvard Club with Charles, who was staying next door at the Algonquin.

The next morning we flew to Bermuda, where our son Robert met us. The four of us checked into our rented unit at Ariel Sands and proceeded to spend a wonderful week on this charming island. We used mopeds to explore. We attended Easter services at St. Peter's Church on St. Georges Island, the oldest continually used Anglican church west of Great Britain. Swimming, tennis, beach walking, moped riding, and shopping at Trimingham's department store consumed most of our days. We also experienced a particularly interesting excursion to the Confederate Museum on the eastern end of the island. During the Civil War, this facility was the headquarters for the blockade-running British and Confederate ships that were supplying the Confederacy's war effort. (Incidentally, during our Bermuda visit, I was reading *Intrepid*, the biography of Sir William Stephenson, the head of MI5, British Intelligence, during World War II. He was living next door to us at the time of our visit.)

On April 10 we flew to Boston. Robert returned to Harvard and we had a visit with Susan, who was working in Cambridge. Pam and I had to return to Santa Fe for school and work, so the two of us flew home while Jeanne diverted to Rome to see Thomas, who had been in Italy for over a year without a visit home.

In June the swallows returned home—Thomas from Italy and Robert from Harvard. Thomas's year and a half of European freedom had increased his strong independence and he made it clear he wanted to be treated as an equal and not as a son. He worked briefly at the agency under Ray's supervision. However, because of his long absence from the University of San Francisco, he had to switch colleges. He enrolled in Loyola University, a Jesuit school in Chicago. He departed in August for his final college year.

Robert spent most of the summer clearing timber at Cow Creek. We visited frequently. His project, and my interest, was to improve game habitat on the property. Jeanne and I would tent on the Cow Creek property but take advantage of the dining room at Martin's Guest Ranch, just a few miles from our land

Back in Santa Fe I continued with my work on the St. John's College board. Dr. Richard Weigle, the long-time president of St. John's, wore two hats as the president of both the Annapolis and the Santa Fe campuses. There was one board of trustees for both colleges, so he was in Santa Fe often. The board met every three months, alternating between campuses. For me, one of the main attractions of being a board member was visiting the Annapolis campus, whose architecture reflected the college's colonial origins. During the build up to the Yorktown Battle in the American Revolution, one of the French units en route to the battle bivouacked on the college grounds. Seeing the contrast between serene St. John's and its bustling neighbor, the U.S. Naval Academy, is a must for visitors. Meetings at Annapolis also permitted me to visit with insurance companies located along the Eastern Seaboard.

A typical trip took place September 17-24. Jeanne and I flew to Boston, where I called on the Commercial Union Company and squeezed in a short visit with Susan in Cambridge. Then Jeanne and I traveled to York Harbor, Maine, for a two-day session with the Kendall General Agency, the insurance underwriter for the agents representing many of the Colorado, New Mexico, and Arizona ski operators. On September 21 we flew to Philadelphia, where I paid a visit to the Insurance Company of North America. Then, a train ride took us to Washington, D.C., where we covered several bases. The Senate was debating natural gas deregulation, an important consideration for our New Mexico gas producers, so we visited our elected representatives, who were involved in this debate. From there, we went to Georgetown University, since Pamela had expressed interest in it as a college potential. I was impressed with its location and emphasis on government and foreign affairs.

Finally we were able to sandwich in a luncheon with our old family friend, Judge Charles Fahy. At the time he was the head judge for the U.S. Court of Appeals. As a young lawyer, he had practiced in Santa Fe. During World War II, he was Deputy Solicitor General and served as the U.S. negotiator for the exchange between the United States and Britain on naval bases for destroyers. Also, I had been in touch with Fahy while I was in Europe after the end of World War II. He set up the Nuremberg Trials for the Nazi war criminals while I was stationed in Austria. Later, my nephew Henry A. Kelly worked as his law clerk in Washington.

In September the St. John's board met at Annapolis, a short drive from Washington. Late Saturday afternoon, after the meeting ended, we flew back to Santa Fe, arriving late at night. By December the ski season was in full swing. Christmas holidays brought the children home, and a productive and generally pleasant year ended.

1978: The Wayfarers

In 1978 our mature children began their individual life-treks along untried trails. For their parents, it was a time of attending graduations and waving adieu as our offspring set off on their ventures.

We went to Chicago in May for Thomas's graduation from Loyola. Afterwards, he returned to Santa Fe and spent two months deciding on his next step. He had thought of joining me in the agency, but his bent towards adventure and his instinctive hesitation about working with and for his father discouraged this move. Instead, he elected to join the Peace Corps, encouraged in part by his contact with David Davenport, Jr., my godson and former Peace Corps volunteer. In August Thomas "enlisted" for a two-year tour and left for orientation in Philadelphia.

On Thomas's flight to Philadelphia, another Peace Corps recruit sat next to him. The two were talking and Thomas noticed his companion was missing the tips of several fingers on his left hand. When he asked about this, his new friend told him that the fingers had been frozen in a climbing accident on Denali the previous summer. On further questioning, "Fingerless" stated that, after hanging upside down in a rope snarl for a freezing night, he was rescued by a group from another expedition attempting to conquer the mountain. He remembered the name of one of his rescuers: Robert Kelly from Santa Fe, New Mexico—Thomas's twin brother.

When he arrived in Philadelphia, Thomas received an assignment to eastern Nepal, where his mission would be to construct a pipeline and water system to a remote village. He spent two years in Nepal as the lone Peace Corps member on the project, which successfully brought water to the village. Thus began Thomas's career in Asia, which is another story to be told.

Shortly after Thomas's graduation Jeanne and I returned to familiar ground in Cambridge, Massachusetts. There, we attended Robert's graduation from Harvard and the thirty-fifth reunion for my class of 1943. We met Robert's friends, he met ours, and we shared a remarkable confluence of similar experiences.

Robert wasted little time hanging around Cambridge. By the end of June, he was off to Peru for a six-month adventure in South and Central America. His college roommate Philippe Bourgois (who, incidentally, became a well-known university professor in social anthropology) joined him for the first part of his adventure. Robert and Philippe floated for a month down the Madre de Dios River in a dugout canoe through the jungles of eastern Peru. They traveled through a region that had little exposure to the outside world, encountering indigenous tribes along the way.

Robert and Philippe parted ways after returning to Lima. Philippe went back to the United States and Robert continued alone, spending the next months in Columbia and Central America. We didn't hear from him until Christmas. He ultimately surfaced in Florida, where the attractive Bokum daughters (formerly of Santa Fe) welcomed him as a guest at their father's estate near Palm Beach. He then returned to Cambridge and launched his artistic career with a job at Polaroid.

Completing the sequence of graduations that spring, Pamela ended her six-year tour at Santa Fe Prep in June. She had had a successful time at Prep, although her athletic ability exceeded her scholastic achievement. She had been a key racer on the ski team and a stalwart player on the field hockey team. Academically, she was adequate but her ambitions for college acceptance did not jibe with the admissions department's demands. Pamela applied to Georgetown, Norwich, Southern Methodist, Tufts, and Harvard. She was accepted at Tufts. We both were disappointed about Harvard but realized Tufts was a fine school.

As a reward for her hard work, we sent Pamela on a trip to Europe. It was her mother's idea. Pam did not want to go. She had a beau, Doug Bell, who she preferred to be with rather than a stodgy tour group. Her mother won and Pam spent six weeks in Europe. In September she began college at Tufts.

Susan, our eldest child, had already left the nest and was working at Crimson Travel in Cambridge. With Robert at Polaroid and Pamela at Tufts, Susan had two siblings close at hand. My business activities led me to the Boston area, permitting me to see them occasionally.

Although Jeanne and I had launched our children, we weren't entirely without dependents: Ranny, now eighty-three, lived next door. She was of an independent nature but she still required our close attention. We invited her often for dinner and included her in activities that appealed to her. She enjoyed seminars at St. John's and events at the School of American Research. Mother also enjoyed spending time with my brother Booker's family, and Yei frequently invited her to family activities in Albuquerque.

The insurance agency continued to be a twenty-four-hour job. Ray Pineda was by now familiar with the routine and had become a minority owner of the business. Undoubtedly, my hard-driving, controlling nature did not always help but, in the main, the business operated smoothly and was profitable. I made my final payment to Gary Noss, retiring the debt I owed him for his interest in the business. This was a great relief and left me debt-free, a situation to which I was unaccustomed.

My nephew Daniel Kelly Gorham, who had graduated from the University of New Mexico, had recently married and needed a job. I knew Dan, and I liked him. He was outgoing and had a good sense of humor and a friendly and warm personality. My junior partner, Ray Pineda, was anxious to hire him. After considerable pondering, I agreed, and Dan joined the business in June and immediately enrolled in a series of insurance schools. In July he had completed his basic training and started what I hoped would be a successful and productive career with the agency.

My Shangri La on Cow Creek was becoming an addiction as a distraction from this insurance business. On weekends, I rushed to the serenity of this little haven of 160 acres within the Santa Fe National Forest, totally obsessed with my project to create out of the heavily timbered wilderness a game sanctuary. I enjoyed the company of the local natives I employed to help me with heavy physical work. I appreciated their uncomplicated characters and the fact that they knew the forest plants and animals. They taught me the ways of life in the woods. I was also becoming proficient in the use of a chain saw, building fences, timber harvesting, and grass sowing.

Jeanne also enjoyed escaping from city life, but living in a tent on weekends was too primitive for her. She did not complain, but obviously missed some creature comforts. This led me to plan some construction projects for later in the summer, but the onset of fall and cold weather delayed construction until the next spring. The arrival of winter also meant it was time for skiing, and the Cow Creek ventures were suspended. The children came home for Christmas, with the exception of Thomas. The wayfarers had returned after completing the first segment of their lifetime journey away from our home.

1979: A Potpourri—Home and Nepal

In January my extended family helped Jeanne celebrate her fifty-seventh birthday. I mixed pleasure with business by taking her to Tucson for a weekend at the charming Arizona Inn. The business end of the combination was an insurance inspection at Mt. Lemmon Ski Valley.

Robert returned home after his six-month backpacking adventure in South and Central America. He spent the winter in Santa Fe fencing property on Tano Road, painting his grandmother's house, and, in spring, he cleared the roads at the Cow Creek property. In June he returned to Boston to commence a career as an artist.

Pamela, at Tufts for her first year of college, spent the spring break in Puerto Rico, and during the summer sojourned with Outward Bound in the Gila Wilderness. In the fall she transferred to Harvard College as a sophomore. Unfortunately, she had missed the important freshman bonding experience at Harvard, but she adapted easily once she completed the transfer.

My next-door neighbor and ward—my mother, Ranny—celebrated her eighty-seventh birthday June 3. Although bright and alert, she was physically quite fragile and required considerable attention from her children. Her brother Francis, "Pip," died in the spring, leaving her as the last survivor of her family. His passing was especially hard for her, as he often visited Ranny in Santa Fe in the summer. She brightened up at Easter and Christmas when the family gathered at her house for dinner. Her old cook Josie came out of the woodwork to prepare the traditional feasts.

Although Thomas had completed his Peace Corps tour, he elected to stay on in Nepal. We decided to visit him on his adopted turf. In October Jeanne, Susan and I flew there. The country was beginning to open up to the outside world, yet it was still rigidly structured and regulated by religious customs and ruled by a God-man king. Thomas had just come out of the bush after two years in eastern Nepal.

We had a joyous reunion at the Kathmandu airport. Thomas was now nearly native. He spoke Nepali and was well acquainted with the culture and geography of the country. This made him the best of guides as he took us around his adopted homeland.

Our first exposure to the capital city was a shocker. The streets were jammed with people, oxcarts, rickshaws, lorries belching black smoke, and countless vendors—all covered with a veneer of grime. We were happy to book in at the Malla Hotel, near the American embassy compound and comfortably removed from the chaos of the city center. The hotel offered us a chance to acclimatize to Nepali life. The U.S. Embassy's western-style cafeteria offered a conventional breakfast, recharging me before I headed out for the adventures of the day.

Thomas took us to Pashupatinath, on the Bagmati River, a sacred tributary of the Ganges. The Shiva temple at Pashupatinath is Nepal's most renowned Hindu cremation site. In accordance with Hindu tradition, a deceased person is wrapped in orange cloth and carried to the temple, where the body is placed on a bier of firewood and burned. (In accordance to strict Hindu rites, the eldest son lights the fire of his father, and the youngest son lights the fire for his mother.) The ashes are then strewn into the river for the voyage down river to the holy Ganges. Visitors to the sacred site also bathe in the sacred Bagmati river, believing this frees them from the cycle of rebirth. Women wash their newly woven saris in its water and then dry them on the riverbank. Apparently, the Nepalese are not concerned about the contamination of water.

Thomas took us on an airplane flight along the great front range of the Himalayas. It was a bright and clear day and Mt. Everest loomed above us at 29,029 feet. The mountains awed us but we were also impressed by the tropical forest in the lowlands, which we also visited. From Kathmandu we flew to the Royal Chitwan Park in the Tarai region. The park is a preserve and is the home of tigers and white rhinos. At the time, the only guest resort in the area was Tiger Tops, which also served as a research center for the study of the endangered wildlife located on the Nepal-India border. The curator of the park had been a faculty member of the University of New Mexico. At Tiger Tops, we explored the jungle on elephants. To mount the beasts, we assembled on an upstairs balcony and transferred to their backs. We took two forays, and spotted both tigers and rhinos from the backs of our trusty transports.

Although we did not experience a mountain trek, Thomas guided us over a large part of Nepal, from Pokara on the west to the Indian border on the east. Short-landing airplanes and Land Rovers allowed us this broad exposure. On our last night we had a Santa Fe reunion at the Visna Guest House in Kathmandu. It was a brand new hotel, not yet opened, built by Ed Bass and his construction crew from Santa Fe. Ed, of the wealthy Bass family, had lived in Santa Fe during his hippy days. He developed and built the La Vereda complex on upper Palace Avenue in Santa Fe. I insured it. Had I thought of it at the time of our Nepalese visit, I could have charged off part of the trip as a business expense for consulting with my client, Mr. Bass.

Our departure from Nepal was bittersweet. We had reestablished a close relationship with Thomas. At the same time we realized he had found his niche in Nepal and had no plans to return to the United States soon. Still, as we departed for Santa Fe we never suspected he would stay in this exotic and volatile part of the world for the next thirty years. (At this writing, he remains there still.)

Our flights to and from Nepal traversed over the Far and Middle East—two parts of the world that took center stage as world events unfolded during 1978. President Carter gained esteem early in the year by initiating and completing the Israel-Egypt Peace Treaty, which was signed by Anwar Sadat and Menachem Begin at Camp David. (Incidentally, my annual paper for the Chile Club that year was titled "Israel, Is It Worth the Risk?")

Despite Carter's achievement, he was not a popular president. The Republican Party was positioning George H.W. Bush as a possible candidate to oppose him. Bush had been active in the early oil play in Midland, Texas. Among his friends were Leland Thompson and Fred Chambers, both of whom had homes in Santa Fe. Owing to their influence, early in the first Bush campaign Barbara Bush was invited to speak at St. John's in Santa Fe. The topic was her experience in China while her husband was there as the U.S. ambassador. I had the pleasure of meeting Barbara when she visited Santa Fe. She was an attractive and powerful woman.

In September George and Barbara returned to Santa Fe, this time to attend a dinner at La Fonda hosted by his Midland connections here. It was a gala affair and well attended. I had the pleasure of dancing with Barbara and meeting George, with whom I was quite impressed.

In Santa Fe, Ed Bennett's offer to purchase the First National Bank was accepted. The old order of local management ended with the arrival of Bennett's Pennsylvania mafia. At the same time, Eberline Instrument Co. was purchased by Thermal Nuclear of New Jersey—a severe blow to the Kelly Agency, as the insurance premium went east with the ownership.

On September 29 fire destroyed the Star Lumber Company, which my agency insured. The total complex—buildings, inventory and equipment—was lost. The company's owner, David Wilson, was a close business friend. Ironically, we had reviewed his insurance coverage a few days before the fire and I had recommended he purchase business interruption insurance. He agreed to consider the coverage, depending on its availability. Insuring an unprotected lumberyard was not a popular risk for insurance companies and, at the time of the fire, I had not been able to find willing insurers. This was a severe blow for Wilson and an agony for me because of the absence of coverage. We paid Star Lumber Company for the destroyed assets but, alas, not for the sales and profit they would have earned.

1980: Cold Wars—Changes in the Ski Insurance Industry

In national politics, 1980 brought the downfall of President Carter. His presidency had begun on a hopeful note as it appeared to be a return to transparency, folksiness, and goodwill after the Nixon-Ford era. As the march of events during his presidency paraded on, however, the Cold War became more frigid and spread to our covert support of Afghanistan against the Russian invasion. In 1980 the botched attempt to rescue American hostages in the Iran embassy, a tragic failure for the United States, caused Carter's approval ratings to collapse. This was the beginning of the end for Carter, who later in the year lost his re-election bid to Ronald Reagan in a conservative landslide.

In some ways the business fortunes of the Kelly Agency followed Carter's. My cold war was fought with competition, particularly in the ski industry. The Western Ski Area Association aggressively marketed an insurance package intended to supplant the monopoly enjoyed by the National Ski Area Association's insurance program, which I marketed. I had previously lost

Santa Fe ski area's insurance coverage but now faced the prospect of further losses.

To match or better the competition, I made two trips to New York and Maine to seek improved coverage and pricing from the underwriters of the national program. I was able to retain Sandia, Angel Fire, Sipapu and Los Alamos ski areas in New Mexico, as well as the three ski areas in Arizona—Sunrise in the White Mountains, Snow Bowl in Flagstaff, and Mt. Lemmon in Tucson.

In addition to these challenges from competition, my agency was responding to big changes in the industry. As computers made marketing procedures more efficient and powerful, the insuring companies were merging and attempting to capture and become the exclusive market for selected independent agencies like mine. I was serenaded by several companies, particularly Aetna Insurance Company and Insurance Company of North America. In April, Jeanne and I were Aetna's guests for a week in Hawaii. We were toured and entertained, in the hope that Aetna would become our main market for life insurance products. In like fashion, the Insurance Company of North America held a one-week seminar for a selected group of agencies it hoped would be their exclusive market for property and casualty business.

This trend for exclusivity did not go unnoticed by alert agencies. It became imperative to modernize and get an overall "makeover" in an effort to be considered as one of the select agencies. I was well aware of this need when I attended the National Association of Insurance Agencies meeting in Phoenix. One of the main speakers, Gary B. Holgate, impressed me with his lecture on agency administration. I engaged him to review the Kelly Agency in a two-day session in Santa Fe. This session led me to improve our operations greatly and got me thinking about how the agency could survive and prosper during this period of change.

Along with the modernization of the agency, I initiated a facelift of my Cow Creek property. I met David Quintana, the fire prevention ranger for the Pecos District of the Santa Fe National Forest. David, a Pecos native, was intimately acquainted with fence building, forest thinning, slash burning, and road construction. The property needed all these improvements, and David was the man to oversee the effort.

On almost every weekend during the summer, I was on location at the property. First, we accurately located the property boundaries and cleared brush and trees to make way for fencing. We packed in via horseback the necessary posts and coils of fencing wire and engaged a crew of young men from Pecos. It was very difficult work as they labored over the steep terrain. By late summer, the crew had pounded in posts and strung four miles of four-strand fencing.

We disposed of the trees and brush we'd removed for the fence lines by burning them in several strategically located burning pits dug by bulldozer operator, Vernon Hutchinson. The pits served well to contain the fires during the dry summer in this thickly timbered area. By the end of the summer, the project was complete. I was pleased to have new fences and gates, and to have the property clear of messy downfall.

While I was stringing fence at Cow Creek, my daughter Pamela was thrashing through the rain forests of Ecuador. She had taken a class in botany at Harvard from a professor who was concerned about the destruction of flora and fauna by oil company exploration in the jungles of eastern Ecuador. He wanted to record and take specimens of endangered plants from this area and sought volunteers for the project. Pamela seized on the opportunity to spend the summer in this remote jungle. There she harvested plants using a machete and lived with indigenous natives in the area. She suffered an injury when one of the workers inadvertently slashed her cheek with a machete stroke. The locals stitched and treated the wound. When Pamela returned in the fall, I noted in my diary that she looked "jungle ravaged."

Upon her return Pam complained about a growing, itchy bump on her shoulder. When she was back at Harvard for the fall semester, she went to the Stillman infirmary to have the swelling examined. The doctor lanced the lump, and out popped a fly larva, almost ready to hatch. This was a first for the Stillman infirmary.

Near the end of the year, a tragedy occurred: on December 24 my long time, grumpy but wise accountant Stan Brumfield committed suicide. A bachelor, Stan lived a solitary existence. He was gruff but had a subtle sense of humor. His clients were his friends, and frustrating the IRS was his delight. He was a wise counselor for Jeanne and me. From a business perspective, losing Stan meant I had to find a new accountant, and I did: Dirk Houtman, who continues to this day to keep me in good stead with the IRS.

Christmas dinner was held at Ranny's house. All the children were home, except for Nepali Thomas. On the last Saturday of the year, the family and dogs hiked San Cristobal Canyon and then drove to attend Mass at the renovated Santa Cruz church. We then went to Jaramillo's Rancho de Chimayó Restaurant for dinner.

1981: Vague Discontentment

This was my twenty-seventh year of managing the insurance agency. At sixty, I was beginning to look up from the grindstone and wonder if I needed to keep up such an ambitious schedule.

I was dealing with clientele temperaments, insurance claim problems, competition yapping at my heels, and personnel problems—and all of it seemed to be mounting in bigger and ever more complex ways. I recognized my attitude was becoming more and more negative, but even though I knew I was part of the problem, I tended to be critical and more demanding of the employees.

The situation came to its head with Ray Pineda. He was in his fifth year at the agency. He was easygoing and affable, but his client base was not stable. His clients were often delinquent in their payments. Instead of encouraging him, I criticized him. This made him unhappy and he decided to leave the agency.

Ray's departure put pressure on Dan Gorham and me to take up the slack. It became obvious that further automating and streamlining the agency operations was necessary. In May, Gary Holgate, a recognized expert on agency management, spent three days examining our operations. He recommended reorganization and more efficient office procedures. Taking his advice, we divided the agency into commercial and personal insurance departments and farmed out accounting. We also became more automated, furnishing each staff member with latest typewriters, which had basic computer capability. The number of insurance companies we represented was reduced. This increased the business volume with the residual companies so the agency became more important and profitable to itself and the companies represented.

I initiated these changes to make the agency more attractive to possible buyers, either individuals or actual insurance companies. Increasingly, I was approached by insurance company field men who sought to associate themselves with independent agencies and become their own masters. At the same time, the national insurance companies were looking for captive agencies to be their exclusive representatives. As the year progressed, this attention began to make me think about what to do with the agency in the long term

As an antidote to the pressures of the agency, I frequently visited the forest refuge at Cow Creek. I engaged George Boylin to install a water system there. George was a unique individual. Fiercely independent and capable—but cranky—he agreed to commute the forty miles of mostly unimproved roads to accomplish the project.

George's three-man crew looked like refugees from a pirate ship. One actually had a hook instead of a hand. Rotund George was always shaded by a fedora, like a gangster. The bulldozer operator was more at home behind the controls of his machine than walking on the ground. Trenching for a four-foot-deep waterline a quarter of a mile up a timbered slope required guts and a maestro's skill, and this crew had it.

The completed water works included the well and a pipeline up the steep slope to an underground storage tank and on to a hydrant located at the main campsite. Having potable water and electricity brought the camp into the twentieth century, making it much more of an attraction for Jeanne than what she had cheerfully endured before.

Along with all the activities that business, housekeeping, and social obligations required, Jeanne was a patient and loving wife and mother, and a good daughter-in-law to Ranny, who was living next door. Now eighty-nine and no longer driving, Ranny required the attention appropriate to someone of her age and position. The obligation added to Jeanne's already full schedule.

Jeanne and I did get away, however, for a change of pace and scenery. We attended a two-day session at a tennis camp near our favorite California lair, La Jolla. We also went to a house party at Frank and Yei's ranch near Phillipsburg, Montana. We were guests along with Malcolm Stewart and his wife Jeannie as well as my college roommate Joe Downer and his wife Janet. Frank, Malcolm, and Joe had been officers in the Tenth Mountain Division in Italy. We passed the time recounting war stories, hiking, horseback riding, trap shooting, and country dancing.

We also engaged in some interesting political discussions, which invariably brought out differences of opinion. As I recall, the topics were the main events of the eventful year. Ronald Reagan had begun his term as president, following Jimmy Carter's bland term. The Iranian embassy hostages were finally released, just moments after Reagan was sworn in. A deranged Turk attempted to assassinate Pope John Paul II, and the assassination of Anwar Sadat threatened the precarious peace between Egypt and Israel. Yei and Frank were great hosts. We had a wonderful time, although I'm sure they were relieved when we finally all departed.

The day after Christmas Jeanne and I visited the mosque at Dar al Islam at Abiquiú. After that, we had hamburgers at La Mesita and ended the day watching the Indian dances at Tesuque Pueblo.

1982: Passing the Baton

My inclination to pass the agency baton to a younger sprinter developed into a reality early in 1982. In February Dan Gorham and I attended an agency management session in Houston, sponsored by the Insurance Company of North America. Founded in 1776, INA was one of the premier insurance companies in the United States. It had grown to be the Kelly Agency's favorite insurance market.

The seminar concentrated on office organization and efficiency. The session also included a strong effort to lure INA's agent representatives into an exclusive contract with INA. The claim was that INA would be the source of all the insurance products an agency would need, which would simplify operations, increase commissions, and help finance ambitious younger agents in acquiring ownership of insurance agencies. It called this program One Compar, meaning "one company participation."

Several agents attending the seminar were already One Compar agents. I found the program appealing and I discussed the pros and cons with several of them. I was particularly impressed with Fred Tiberi, who owned a relatively new agency in Albuquerque. Fred was enthusiastic about the Compar program. When I returned to Santa Fe, I sought him out to further investigate the program. Our relationship developed as I picked his brain, and he became interested in acquiring the Kelly Agency.

Fred was in his early thirties. He was originally from Chicago and attended the University of New Mexico, aided by a tennis scholarship. After graduation he had been trained and employed by a major insurance company. With this background he later established the Tiberi Insurance Agency in Albuquerque and was a One Compar INA agent. He was personable, bright, ambitious, and an excellent tennis player. In March he presented me with a proposition to buy the Kelly agency with financing from INA. The offer was to split ownership, with Dan Gorham owning 20 percent as a junior partner.

The proposition had merit. For the next several months I maintained normal operation of the agency while I sought advice and counsel from experts about the proposed deal. I engaged David Hales from Chicago as my expert advisor. Meanwhile, the INA initiated a thorough audit of the Kelly Agency's records, finances and its book of business. We reached a mutually agreeable purchase price. On August 31 the purchase agreement was signed, effective October 1, 1982.

As these events unfolded, our normal business and personal lives continued. Jeanne commented that we lived our lives like squirrels in a cage, with a lot of repetition in the routine. However, there were important events. In June we attended Pamela's graduation from Harvard. Mother Theresa was the featured speaker. From there, we visited Martha's Vineyard. (I remember on this trip crossing the Chappaquiddick Bridge of Teddy Kennedy's infamy.)

On the international stage, the British defeated Argentina in the Falkland Islands war while, at home, our friend Bob Turner was in a terrible automobile accident that left him a paraplegic—yet he remained a courageous example of a continually involved citizen. Another tragic event was the brutal murder (still unsolved) of Franciscan Father Reynaldo Rivera, the rector of St. Francis Cathedral. Closer to home, our family cook and devoted servant, Josefita Gorman, died, as did Marion Dockweiler, the charming tenant of our guest house. (Marion's family was the longtime owner and operator of the Cow Creek guest ranch.)

Finally the epic day of October first arrived—the day I would divest myself of the insurance agency that had been a major focus of my life for thirty-some years. The Kelly Agency was now owned by Fred Tiberi and Dan Gorham. Don Van Soelen of the First National Bank acknowledged receipt of a wire transfer from INA in payment. I now was no longer the owner but rather an employee of the Kelly Agency. As part of the terms of the purchase, and as a guarantee for the financial backing of INA, I was obligated to actively serve as an advisor and consultant to the agency for several years. The deal guaranteed the perpetuation of the agency and an adequate payoff for me. My responsibility remained to help insure the successful and profitable continuation of the agency.

Bud and Jeanne Spooning, ca. 1949

Jeanne's Announcement Photo, 1953

The Wedding Party, New York City, 28 November 1953

Bud and Jeanne's Wedding Portrait

On Honeymoon, Tijuana, Mexico, 1953

Robert, Susan, and Thomas, "Pajama Party," 1958

Jeanne and Kids at Guaymas, Mexico, 1963

Bud, Tom, and Bob, Santa Fe Ski Area, 1965

Jeanne with Twins in Easter Finery, 1963

1983-1993

TRANSITIONS AND FAREWELLS

1983: The New and the Old

Since I remained actively involved with the insurance agency but no longer held the leadership position, I was busy but without ultimate responsibility. This meant there was less pressure on me. At the same time, though, I subconsciously felt resentment over loss of control.

My relationships with Fred Tiberi and Dan Gorham differed. Fred was the obvious boss. He had considerably more management experience than Dan. He was pleasantly laid back, amiable, thoughtful, and diplomatic in his relations with me, and he recognized my importance in dealing with our major accounts. As a result, we worked together harmoniously. Although Dan had worked for me for a while, he was relatively new to the industry. Now an owner, he may have felt somewhat uncomfortable in his new relationship to me and the business. Both he and Fred had a calmer and more tolerant attitude toward the staff than I did. This was a decided improvement.

My involvement with the agency was like that of a pilot guiding an airplane on a gradual slow descent to a safe landing. I still had business relationships that allowed Jeanne and me to travel often in a mix of pleasure and business to many places, including the Sunrise and Mt. Lemmon ski areas in Arizona and trustee meetings for St. John's College in Annapolis. The Annapolis trips allowed Jeanne to visit her eastern relatives and friends, and the June 1983 meeting also conveniently coincided with my fortieth college reunion in Cambridge. In one trip, we were able to rendezvous with Robert and Susan and visit my college classmates on my way to meet with insurance companies in Boston.

In April we commemorated the tenth anniversary of the death of Tom Old, our longtime friend. At the Kiva Club, members reverently opened Tom's locker and solemnly drank his twenty-year-old Scotch, ten years after the airplane crash in Alaska that killed him. Another important figure departed the stage in August—Santa Fe architect John Gaw Meem, a leading figure in defining the historic character of Santa Fe.

On the national scene, the Marine Corps barracks in Beirut was bombed in October, with the loss of over one hundred men. Only a couple of days later, the United States invaded Granada, presumably to rescue students endangered by a Communist-inspired coup. William Zeckendorf and St. John's College jointly developed the gated Los Miradores condominium complex. Burch Ault, the college provost, did a masterful job in negotiating the deal.

We celebrated our thirtieth wedding anniversary at home on November 27. The happy event included toasts, dancing and, the next day, hangovers for our family and friends.

At year's end, Thomas was in western Nepal; Robert was painting in Italy; Pamela was home, in transition; and Susan was working in real estate with Ralph Brutsche, a developer of upscale housing in Santa Fe. My journal notes, "Jeanne and I have an interesting and comfortable existence. We need challenges and interests. We are really pretty young and need to be vital—look to 1984."

1984: Interludes in the Routine

Very little change occurred in my day-to-day activity during 1984, the second year of my so-called retirement from active management of the agency. Fred and Dan, the new owners, required briefing about the accounts and the principal individuals involved. Recognizing my value in this area, both were anxious to accommodate me. Perks and the challenge of the work furnished me with a strong incentive to continue.

I worked mostly with Fred, the senior partner. He accompanied me to Annapolis for two St. John's College meetings and then on to visit insurance companies in New York City, Boston, and Philadelphia, to keep current with changes in the industry. Furthermore, since the Insurance Company of North

America was footing the loan used to purchase the agency, my presence at its loan review meetings reassured the lender.

Jeanne and I enjoyed two trips during the year. In early April we went to Cat Cay in the Bahamas, where we were guests of our friends the Dillons. The Bahamas had recently severed colonial ties to Britain to become a sovereign island nation. Cat Key, a small island only thirty miles off the Florida coast, served as home base for some of the racketeers who operated in Florida and the Bahamas. It had also been the secret trysting place for King Edward VII and his femme, Wallis Warfield Simpson, prior to his abdication from the throne. We enjoyed nearly unoccupied beaches, snorkeled amid colorful coral, swam, played golf and tennis, drank rum, and generally relaxed and unwound for a week.

In October we joined Malcolm and Rosie Stewart for a cruise down the Danube River from Vienna to the Black Sea and then on to Istanbul, Turkey. I had not been in Vienna since 1946 when I was in the army. We stayed at the Hotel Imperial, a sumptuous palace that had been the headquarters for the Russian KGB during my wartime stay. Then we boarded the river ship, MS *Moldavia*, to ply the Danube River to the Black Sea. The top deck of the clean, comfortable ship offered a clear view of both banks of the river. Since the ship was operated by Russians, the vodka and caviar were excellent.

With the Iron Curtain still in place, Russian troops and Communist regimes controlled the several countries we floated through. The ship docked at ports on the river, where we took buses into the countryside and cities. The voyage took us to Slovakia; Budapest, Hungary; Belgrade, Yugoslavia; Bulgaria, Romania, and the Russian enclave, Izmail, at the mouth of the Danube. From our vantage point as free passengers we witnessed and experienced, to a small degree, the regimentation and subjugation of these countries under the yoke of Soviet Communism.

I remember visiting one ancient church where several old women were gathered. When they saw Jeanne with a rosary in her hand, they surrounded her with much interest and affection. It was obvious they longed for the religious freedom that Jeanne obviously enjoyed.

The climax of the trip was the voyage from Izmail into the Black Sea and on through the Bosporus and Golden Horn to Istanbul, Turkey, on the Russian ship *Ayvazouskiy*, which, compared to the *Moldavia*, was a derelict. After three days in Istanbul, we flew back to the United States and home to New Mexico.

These trips provided Jeanne and me with important interludes in the routine of our lives. Our relationship was on solid ground, but my tendency to be in perpetual motion and the persistent worries of work and home life kept the tension high. Perhaps this was characteristic of my generation of men, as a number of our friends were having marital problems.

My college roommate Joe Downer and his wife were splitting up. Joe was vice chairman of the board of Atlantic Richfield Oil Co. and a past president of the Harvard College alumni board. In all respects he represented the ideal of the successful corporate executive, but his hard-driving habit had a down side. When his wife Jeannie had had enough, she departed. The blow to Joe's ego and to his corporate image was shattering. In a state of shock, Joe sought refuge with us. We did our best to help him recover during two separate stays, introducing him to friends who had successfully overcome similar breakups. We entertained him, listened to him, advised him. Ultimately, he recovered and later remarried, but seeing him go through the divorce was a lesson for me, to the benefit of my relationship with Jeanne.

When we returned from the Danube cruise, my mother was in bad shape. She had suffered a minor stroke during our absence. She recovered, but a more severe stroke hit her shortly after our return, requiring hospitalization and leaving her impaired and in need for almost constant care. Nursing homes were not yet widely available, so the decision was made to keep her in her own home. Fortunately, we found an experienced nurse, Stella Lobato, to serve as Ranny's primary caregiver. Stella was saintly in her attentiveness to Mother, radiating great devotion and warmth. Mother cherished her. Stella kept a nightly vigil while Jeanne and I and other members of the family handled the daytime duties.

Mother was not a patient invalid. She was restless, frustrated, and could be irritable. This was taxing for me, as I was not a compassionate caregiver by nature. I often regretted my lack of empathy and patience and tried to make up for my impatience by taking her on drives or by reading to her. Jeanne, though much more thoughtful and considerate, also found this constant caregiving regime stressful and difficult. Fortunately Mother improved during the Christmas season. The return of her grandchildren for the holidays and the general festive atmosphere of the season lifted her spirits. On Christmas Day dinner was served in Mother's dining room, as in earlier days. She sat at her usual head-of-table position and presided over the feast. She was soon to be ninety-three and the sand in her hourglass was about gone, but it was a beautiful Christmas Day to remember.

1985: Mother's Passing

For the first months of 1985, Mother's slow and torturous dying required a daily watch at her bedside. Stella bore the main burden of this vigil, but Jeanne and I filled in to give her release. These were agonizing times. Poor mother simply lay prone, her mind undoubtedly tormented by the frustrations of a nonfunctioning body. When not dozing, her eyes reflected the purgatory she was enduring.

Relief finally came when she died on May 17, one month short of ninety-three years. At her funeral at St. Francis Cathedral, Fr. Anderson Bakewell, a family friend, conducted the service. Her ten grandsons, all of whom she had cherished, served as pallbearers. Robert, her artist grandson, read her favorite poem, "On the Road to Cundiyo," by Alice Corbin Henderson. Mother was interred next to Dad in the family plot at Rosario Cemetery. The day ended with a gathering at her home of sixty years at 531 Palace Avenue.

I was the executor of Mother's estate and had to launch immediately into that role. It took considerable time to arrange for appraisals of personal property, valuation of the home, and to publish notifications and distribute her belongings. My attorney brother, Booker, had outlined these necessary steps before Mother's death, but unfortunately he departed shortly before she died for a six-month sabbatical in Europe, which left the legal details of the estate settlement up to me.

Adding to the boiling of the estate pot was Booker's sudden and severe illness in Germany. He developed a life-threatening infection in a heart valve. His wife, Susan, had to transport him hurriedly from Germany across France to the nearest hospital in England. Doctors there knew his history of heart problems and diagnosed the problem. However, the solution was not simple; it required replacement of the infected heart valve and ridding the rest of his system of the deadly infection.

Recognizing the gravity of his condition, and the possibility he might not survive the operation, Booker was adamant he have the surgery at home. Arranging for a flight from England to Albuquerque presented a major obstacle because airlines prohibited non-ambulatory passengers from flying. Furthermore, he would require injections en route to keep the infection from rampaging. With the help of Booker's Santa Fe physician, Dr. Streeper, and with the cooperation of the British doctors, we were able to schedule injections in Britain and St.

Louis en route to Albuquerque and the hospital. All the while we had to keep Booker looking "ambulatory."

In Albuquerque Booker was immediately placed in intensive care. The successful operation took place on July 31 and he finally was permitted to return to normal living on September 4. His race across the ocean and his operation and subsequent time of recuperation added trauma to the whole family's life.

Another activity contributed to the drama of this marathon year. As a trustee of St. John's College, I was appointed to a search committee to select a new college president. The board normally met every three months, alternating between the campuses at Annapolis and Santa Fe. In between the regular meetings, the search committee met either in Chicago or New York. The end result was that I attended seven different meetings during the year, each one lasting at least two days. At our final meeting for the year, we interviewed George McGovern, a former U.S. presidential candidate, but we didn't make a definitive choice.

In August I built a new access road at Cow Creek. Prior to this, a visit to our favorite camping spot required traversing U.S. Forest Service property. The new road would be entirely on my property, with the exception of a one-hundred-foot easement from the Forest Service. The road would climb approximately three hundred feet in about one-half mile and had to be cut through a forested area of mature trees interspersed with several large rocks.

Fortunately, my foreman, retired forester Dave Quintana, was an experienced sapper and knew how to use dynamite to fracture and splinter large boulders. I had to obtain the necessary dynamite and explosive caps to do the job. Through Dave's connections, I bought the explosives in Albuquerque and gingerly conveyed them up to the construction site, a distance of about 140 miles.

Prior to the rock removal, the road route had to be cleared of timber. The timber crew, of which I was a member, chain sawed the trees from the right of way. A contract bulldozer operator leveled the road and pulled the tree stumps from the route. On the day of the blasting, August 13, we posted guards on the little-used Cow Creek public road. Dave positioned the dynamite on the unforgiving rocks. With the caps placed and a long fuse snaking to the boulders, we blasted two rocks about the size of large automobiles into workable pieces. The route had been cleared, and the new road was opened to four-wheel-drive vehicles. This road still well serves its intended purpose.

In late September we were fortunate to join John and Sis Dillon for a safari in Kenya. The Dillons were old Africa hands, having experienced two prior trips. Our three-week safari, mostly in a simple van with driver, covered a large part of Kenya and most of the game parks.

Upon arriving in the capital, Nairobi, we flew to Malindi on the coast of the Indian Ocean and drove sixty miles inland to the Galana Ranch, a privately owned empire over a million acres in size. In addition to being home to a large and varied wild game population, it was a cattle ranch serving east African beef markets centered in Mombassa. There were few restrictions on game viewing on this private ranch. Manager Henry Henley, who had been a "white hunter," drove us in a Jeep, rounding up groups of elephants and generally arousing the wild game population, much like a cattle roundup in New Mexico. It was a real thriller.

Our first night at Galana, Henley received a radio message from his northern foreman, some forty miles away: two young male lions were killing cattle. The foreman requested help. I volunteered to go to the troubled area with Henley and his helper James. After an hour drive in the Stygian darkness, we arrived in the vicinity of the kills. We walked past a large herd of cattle guarded by several Somali herdsmen and ensconced ourselves in a thorn bush enclosure some two-hundred yards from a cattle carcass. Henry was armed with his lion rifle, and James was equipped with a flashlight that illuminated the lion bait. The night was black except for the bright stars above. On two occasions, James threw his torch light onto the carcass, only to disturb hyenas tearing at the dead cow. The wily lions did not attack the bait. We returned to the ranch lodge early in the morning grumpy and unrequited. (Henley had a storeroom filled with lion skins from more rewarding stakeouts.)

While in the Rift Valley at Lake Naivasha, we visited Joan Root, a renowned naturalist. Her father, ex-Kenyan Edmund Thorpe, was our friend in Santa Fe. Joan lived alone and wore a pistol. At night, hippos wandered from the lake to her garden area. We had a delightful afternoon with her. (Some twenty years later, Jane was murdered by a local, despite her efforts to defend herself.)

In the Masai Mara Game Reserve, we took an early morning balloon ride, floating over a large stretch of river to look down on hippos, crocs, and game at watering holes. This region was the home of the magnificently handsome and leggy Masai herdsmen, noted for their walking speed and endurance.

During our stay in Kenya, the movie *Out of Africa* was the hit. On our final day there, we had an emotional thrill flying into Nairobi over the Ngong Hills, the setting of the last scene in the movie.

Once back in Santa Fe, events leveled out and the usual routine set in. Christmas arrived. The year ended peacefully.

1986: The Phoenix

St. John College's search for new leadership to replace President Edward de Lattre extended well into 1986. One reason for the prolonged search was the policy change from having one president for both Annapolis and Santa Fe to having a president at each campus. The faculties of both campuses had their own ideas about who their president should be. The alumni were deeply involved in the ramifications of this policy change. The board of governors was responsive to those pressures, but we had to make decisions based on best interests of the college as a whole.

We had been meeting for some time and had considered George McGovern for one of the posts. The Annapolis faculty admired McGovern's national reputation, but members of the board felt otherwise. The Santa Fe faculty was somewhat ambivalent. In the end, we passed over McGovern and selected William Dyal for Annapolis and Michael P. Riccards for Santa Fe, initiating the change from having one omnipotent president to having two independent but cooperating presidents answering to a single board of trustees.

Following a January 2 meeting of the search committee in New York City, I returned to Santa Fe on January 6, where Jeanne, Pamela and Robert met me with devastating news: the Kelly agency offices, along with the rest of the Padre Gallegos building, had been destroyed by fire. The upscale Santacafé restaurant and the Kelly Agency offices were separated by a common masonry wall, against which the café's cooking range was positioned. My office was adjacent to this wall. There were no firewalls in the space above the ceilings to isolate the rooms. During the night, a grease fire ignited in the restaurant kitchen and the flames burned through the ceiling area and spread through the entire building.

The enflamed roof fell into my office and destroyed everything. Fortunately, no customer files were located there, but paintings, rugs and other irreplaceable items were consumed. Perhaps the greatest personal loss was several minute books of Otero, Sellar & Co. and its successor Gross Kelly & Co., which covered over one hundred years of Southwestern mercantile history.

Fred, Dan, and the shocked staff heroically salvaged the working vitals of the agency. The adjacent McKee building was partially empty, and our agency moved into this building within two days of the fire. Business was not "as usual," but the agency was effectively functional.

The Padre Gallegos building was a historic structure. Prior to and following General Stephen W. Kearny's conquest of New Mexico, it was the home of Padre José Manuel Gallegos, an educator, printer, and the religious connection between the bishop in Mexico and his isolated Catholic flock in northern New Mexico. Gallegos's residence was the intellectual center of Santa Fe. Following the arrival of the first U.S. Bishop, Jean-Baptiste Lamy, the padre lost his influence and his residence fell into disrepair. It was in shambles until restored by R. E. McKee Construction Co. in the 1970s.

Fortunately for the Kelly Agency, McKee spared no time or cost in restoring the building following the fire. In November the agency moved back into its spacious and attractively efficient home.

During part of the reconstruction exile, Jeanne and I visited West Germany where we were joined by my sister Yei and her husband Frank. The tour started and ended in Brussels. The Iron Curtain was still rigidly in place between the two Germanys, which made the trip also very interesting politically. During World War II, Frank Gorham had served in Italy with the Tenth Mountain Division fighting against the Germans, and I had served in France and Germany. In a sense, returning there was a sentimental journey.

Our tour initially went south through Bavaria and into northern, Tyrolean Italy, an area more German than Italian. While staying at Schloss Labers, the Chernobyl nuclear reactor meltdown happened in the Russian Ukraine. There was concern that the resulting nuclear cloud, driven by prevailing winds over northern Italy, would contaminate fresh food supplies As a result, consumption of milk, fresh vegetables and fruit was prohibited, putting an unfortunate crimp in our meals at this charmingly peaceful part of the world. While at Schloss Labers, we accompanied ex-Captain Gorham south to the Lago di Garda area where some forty-one years before the German army in Italy finally surrendered, essentially to the Tenth Mountain Division.

Back in Germany, much of our route paralleled the East German border. This political division was forcibly evident. The route followed a continuous cleared area, complete with a cement apron and a high fence topped with manned observation towers. We felt blessed to be on the outside of this prison.

We ended our tour in Brussels, Belgium, the headquarters of NATO. Upon arriving, we feared there was a political uprising occurring because automobile horns were blowing and crowds of shouting people lined the streets. Instead of a rebellion, though, it was a glorious celebration welcoming home the national soccer team, which had earned second place in the trans-European soccer matches.

On our return, Santa Fe seemed small, friendly, and comfortable. As the year ended, I contemplated the challenge ahead: profitably selling my mother's home at 531 Palace Avenue and amicably settling her estate with my siblings.

1987: Changes in the Air

Change was in the air in 1987. The children were in and out of Santa Fe during the year. Pamela had been a legislative assistant for Senator Bingaman for four years and was growing restive with the Washington political scene. She applied for the MBA program at New York University and was accepted. However, the lure of California clouded her thinking and, instead, she chose the program offered by the Santa Clara University. I was upset, as I felt New York University was the better school. In any case, after a free summer, she commenced her new role as a graduate student in September.

Thomas, now a ten-year expatriate in Nepal, made several trips to the United States. Abbeville Press had published his first book, *The Hidden Himalayas*. It had been relatively well received and he readily accepted the lecture invitations that came his way. Two of the invitations were in New Mexico, in Santa Fe and Los Alamos. On these occasions, he was a welcomed visitor back in his old home turf. Robert, the painter/artist, was now a permanent New Yorker and had several gallery shows there during the year, as well as in Texas, California, and New Mexico, which allowed him also to visit Santa Fe. Susan, an agent with Barker Realty in Santa Fe, remained active in the community and kept in close touch with Jeanne and me.

We enjoyed the children's company and they enjoyed ours. Our reunions were infrequent, but when they did occur, our close-knit relationships were reaffirmed. Taking them to visit Cow Creek was always a must. I was anxious to show them the progress I had made in cleaning up and improving the timber stands during my weekend sojourns there.

While en route to Cow Creek on July 3, I stopped at the old Pigeon Ranch adobe, the primary site of combat during the Civil War battle of Glorieta in March 1862. I had heard a Confederate gravesite had been discovered and was being excavated. My timing was perfect: the topsoil had just been removed and the gravesite opened. In it lay the remains of Confederate soldiers killed that cold day 125 years before. With military precision, the uniformed skeletons lay aligned in parallel rows in the square grave. The men had been buried in their battle dress. Remnants of leather shirt cuffs and neckbands hung on the skeletons, and boots or the soles of shoes were still protecting the foot bones pointed skyward. It set my heart pounding to see this array of combat dead who were obviously buried with dignity and emotional recognition by their comrades. (Some time later a federal act was passed legalizing the reburial of these "rebels" in the National Cemetery in Santa Fe.)

As executor of Mother's estate, I was involved in the renovation and care of her house and property at 531 Palace Avenue. The heirs had determined the property should be sold at a proper price to a buyer compatible and empathetic with the emotional legatees concerned. This was a challenging real estate problem for me. Despite—or perhaps because of—all these responsibilities, Jeanne and I decided to get away for a trip. On June 1 we flew to Portugal and boarded a cruise ship that would ultimately dock at Bergen, Norway, after sailing to Spain, France, Ireland, and the outer islands of Scotland. It was here in the Orkney Islands that the German navy was impounded in the Scapa Flow at the end of World War I. Our trip ended with a whirlwind, two-day tour of the fjords of Norway.

When we returned to Santa Fe, I continued at the agency with my duties and responsibilities as senior counsel. As the year went by, my time was increasingly involved in mediation between the two new owners of the Kelly Agency, Dan Gorham and Fred Tiberi. In late June the long-simmering friction between Gorham and Tiberi became a major distraction for business operation. Tiberi determined unequivocally that he wanted to dissolve the partnership, but Dan resisted. The situation became a legal stalemate, and I found myself in the uncomfortable middle of an untenable situation. Dan was my nephew and former employee. Fred was the senior partner whom I admired and had worked with closely for over two years. I struggled for some kind of compromise.

1988: The Conflict

The rift between Fred Tiberi and Dan Gorham reached a climax in January 1988. Despite my best effort to effect a reconciliation, the rift was beyond repair. On January 7 Fred held a stockholders meeting. He was the only director and stockholder present. Dan was, in essence, terminated. After the stockholder meeting, Fred called a staff meeting and announced Dan no longer had the authority to transact business.

A suit was filed and a preliminary court hearing ruled Dan's termination was done in accordance with proper procedures. However, a full hearing to determine the legality of the dismissal was scheduled for April.

I was neither an officer nor a stockholder, yet I was under contract to continue my involvement in managing the agency. Because the legal preparations for the trial absorbed most of Fred's attention, my duties and responsibilities for steering the operation of the agency increased substantially.

The parties reached a settlement in March. It finalized the "divorce" but required a substantial payment to Dan. Fred was now the sole owner of the agency.

Meanwhile, I was working to resolve the disposition of Mother's Palace Avenue property. It had been three years since she died and the estate still had not been settled. The primary asset was the property but, as yet, I had received no reasonable purchase offer. My brother Booker and I considered purchasing the guesthouse separately. Booker was interested in acquiring it for his married daughter, Rachel, while I was interested in owning it for my son Thomas, who was living precariously in Nepal and might return. Determining a reasonable price for the guest house was complicated because of its proximity to the main house. The matter was finally settled when both of us withdrew our purchase offers. The disposition of the guesthouse became integral to the sale of the property as a whole.

In July we received a reasonable offer from Robert Newfeld of Houston. The deal closed on July 21. Following necessary payments and adjustments, each heir received payment. At long last, the final accounting was done and the estate was closed in November. My three years as executor finally came to a welcomed ending.

For the first time, a stranger owned the property next door. After sixty-five years, the compound was now split; there was no ownership connection between 531-533 Palace Avenue and our home at 535.

In between the "divorce" at the agency and the estate settlement, Jeanne and I attended my forty-fifth Harvard class reunion. When we returned home to Santa Fe, Jeanne began feeling strangely. She had developed a rash and was experiencing discomfort in other ways as well. In September she detected a lump on her breast. A physical exam suggested the possibility of breast cancer, and a biopsy confirmed our worst fears: she had a cancerous tumor. She had to decide whether to have a lymph gland operation and chemotherapy or breast removal. This was quite a shock. What a decision to face.

On September 22 we had a session with Dr. Grey Gordon. The prognosis didn't look good. The cancer was active and was not limited to the breast. He presented the treatment options in such a clinical and cold manner that we were completely wilted and overcome. A consultation with our retired friend Dr. Weebe led us to choose the option of surgical removal of the cancerous breast.

On September 30 the surgeon, Dr. Seedman, performed the operation. Jeanne was released from the hospital on October 3 and returned home. A new life commenced, made livable by the attention and devotion of Jeanne's many close friends.

On October 12 we flew to Dallas for a consultation with a chemo-radiology specialist, Dr. Jones. He recommended a treatment cycle that began on our return under the supervision of Dr. Greenfield. For the balance of the year, Jeanne dealt courageously with the chemotherapy cycles while maintaining an active and mostly normal life. Christmas brought the children home as Jeanne's treatment continued. The turbulent year ended with the family emotionally scarred but optimistic.

1989: Cow Creek and the Erection Party

Although my tenure at the Kelly Agency was supposed to fade as the new ownership gained experience and maturity, in early 1989 I seemed to be reverting to my center-stage role. Fred Tiberi, now the sole owner, was the victim of two blows that altered the course the agency was taking.

First, the ambitious renovation of the office after the fire had been more costly than anticipated. The remodeled structure was much larger, and although the layout of offices and their furnishings was an attractive melding of efficiency, comfort and decoration, the facility was more elegant and expensive than the business justified. The cost added substantially to the debt Fred had assumed when he purchased the agency.

The second blow came from the settlement between Fred and Dan, which left Fred owing Dan. Furthermore, Fred was ambitious. He was anxious to become part of the new health insurance market that was developing. He became involved in a new venture aimed at changing the way medical doctor associations were reimbursed for fees. This new endeavor diluted Fred's attention from managing the Kelly Agency. He delegated much of the responsibility for running the office to Joanne Legits, a veteran of some twenty-two years with the agency. Although Joanne was eminently capable, she lacked the empathy and patience to orchestrate a smooth running office staff.

This left that responsibility to Fred, but it increasingly fell to me, since his attention was diverted elsewhere. I took this assignment seriously, but I lacked the intense dedication I had previously felt when I was an active owner. As a retrogressive retiree, I did not seek additional responsibilities. The thought of total retirement became increasingly attractive.

To further complicate matters, Fred's health deteriorated, twice requiring hospitalization. Competition was severe, and we lost a number of key accounts. Servicing the debt became a problem. I advanced Fred three short-term loans during the year to cover cash flow deficiencies.

As we worked to retrench and cut costs to combat the competition, Fred underwent surgery. I recognized Fred's problems. I liked him, and he was always extremely considerate and gracious to me. I felt an obligation to help him and did so, but at the same time I had sold the business in order to free up my life. Despite Fred's problems and my commitment to be of assistance, I made it a point to maintain my outside interests and obligations during these difficult times for the agency.

My property on Cow Creek was my weekend psychiatric couch. There were no insurance problems there, no board meetings—only nature and property maintenance requiring attention. I was addicted to the physical exertion of working there, assisted by a crew of friendly and capable locals.

Over the decades, the property of 160 acres surrounded by the Santa Fe National Forest had become overgrown. Forest debris littered the few open spaces and trees were jammed so close together that proper growth was impeded. Nature had overdone its job. The need for open space and sunlight was critical. My challenge was to rescue the area from nature's excess. Fortunately, the Valencia family came through like heroes.

The Valencias were patricians. Originally from the lower Rio Grande Valley, they homesteaded in the Pecos area during the Mexican era. Fermin, the patron, was a rancher, a *curandero* (healer), and an *hermano major* (elder brother in the Penitente brotherhood). The necessity for cash income required him to work in Santa Fe at La Fonda Hotel, where he did scullery work in the hotel kitchen. On weekends, in contrast, he was a leading citizen and landowner in rural Pecos, a respectable, dignified and wise man.

Fermin's sons, George, Eloy and Gary, along with their cousin David Quintana, were my weekend warriors, combating the forest evils on my property. Their mother's maiden name was Bowles. She was thoroughly Hispanic, but her great-great-grandfather was a Virginian named Beverly Bowles. As a young soldier, Bowles came through Pecos with General Kearney's army. Later, after the Civil War, he returned to Pecos, married and was the genesis of the blond-headed Bowles family that became well established in the Pecos area.

The main achievement of the Cow Creek crew this year was the construction of a tent platform and a separate latrine and shower house. Jeanne had suffered for years sleeping in a dirt-floored tent, cooking on an outdoor fireplace, and performing required ablutions in a primitive lean-to. I felt the improvements were the least I could do to enhance her comfort when she joined me on occasional weekends at Cow Creek.

In late August the construction project was complete. The all-weather, African safari-like tent, fourteen feet high and twenty feet square, rested on a raised wooden deck. It was equipped with electricity, a stove, and a nearby water hydrant. In near proximity was a wood-frame latrine with an attached shower room and water sink—almost all the comforts of home. The formal tent raising was a special event, attended by the Valencia wives and children. In subsequent years, the "erection party" became an annual summer fiesta day to raise the tent, which was removed each winter.

Jeanne and I enjoyed several trips, in part in deference to Jeanne's recovery regime and other health problems she endured during the year. Our first jaunt took place in February. At an opera fundraiser, we won at auction a trip to New York City. We stayed for a week in a charming apartment owned by the Rutgers Barclays at 169 East Seventy-ninth Street, adjacent to Central Park. While there, we were able to visit our son Robert, a working New York artist. The week was an excellent boost to Jeanne's morale.

In April, Jeanne accompanied me on a trip to Arizona, where I made my insurance calls on the Arizona ski runs. We went on to stay for a week in La Jolla, California, at the Beach and Tennis Club where, thirty-six years before, we had enjoyed our honeymoon. In August we returned to California to attend

Pamela's graduation at the University of Santa Clara. Her Master's degree was in Agricultural Economics, but I felt her real major was the friendships she made. We met Pamela's classmates and toured the fabled city of San Francisco.

Our big trip for the year was a tour of Northern Italy in September. We traveled with the Malcolm Stewarts and David Wilhelms from Denver. With Malcolm driving, we visited Milan, Venice, Osola, Padua, Bologna, Florence, Lucca, Genoa, and Lake Como. One of the highlights of the trip was Malcolm's guided tour of the famed Riva Ridge, where his Tenth Mountain Division assaulted and captured the high ground, exposing the flank of the German army on Mt. Belvedere. This forced the Germans into the open Po Valley and, ultimately, to surrender.

The climax of the Italian tour was a plunge into luxury. We spent three days at the famed Villa d'Este Hotel on Lake Como, a "pleasure dome" offering everything to satisfy the "B.P." (beautiful people) who congregate at this idyllic complex in season.

After the Villa d'Este splurge, I had to return to Santa Fe to recoup the costs. Reality returned. Jeanne had another operation to examine a spot on her lung. This proved to be a benign granuloma, rather than a cancerous growth, but the operation was another tortuous experience for her and caused much anguish to the family.

In world affairs, it was a suspense-filled year. George Bush was elected president, Japan's Emperor Hirohito died, and Gorbachev visited the U.S. The Exxon-Valdez oil spill disaster unfolded in Alaska, and San Francisco was ravaged by a severe earthquake. In December the U.S. Army invaded Panama to oust the corrupt dictator Manuel Noriega.

In spite of the gravity of events in our family and in the world, at year's end I made this optimistic entry in my diary: "The last months of this year have seen the crumbling of the Soviet's Eastern European Communist block. The whole focus has changed. The Cold War seems over and Russia is a patient and not a threat."

1990: The Revolving Stage

Our adult children monopolized the stage in 1990. Although they were globally scattered and involved in a variety of career experiments, they entered and departed from Santa Fe in theatrical appearances throughout the year.

Susan took the lead role in January and February. She had been in a one-car accident in late December as she was returning home solo from a party. She lost control of her car and ended up in the middle of a large juniper tree off of Old Taos Highway. Fortunately, her tank-like Saab was not badly damaged. Susan, however, suffered a broken wrist and was badly bruised from the impact and scratched by broken glass and juniper branches. She was hospitalized for several days at St. Vincent Hospital.

To her credit, Susan recognized the cause of the accident and enrolled in a chemical dependency course at the hospital. The treatment regimen required parental participation. During the course of the program, I recognized my impatience and temper intimidated the children when they were young and that this could have contributed to Susan's difficulties. This realization prompted me to join Al-Anon. Meetings there gave me insight into my own dependency problems. Susan made a full recovery and went on to graduate with honors.

Thomas and his wife Carroll Dunham were both employed by the English natural cosmetic company known as The Body Shop. The company, which was founded by Gordon and Anita Roddick, collected natural herbs from around the world and, after processing them, marketed natural beauty products at Body Shop stores worldwide.

Thomas and Carroll had met years before in Nepal. Carroll had graduated from Princeton in 1985, summa cum laude. Her field of concentration was Anthropology, with an emphasis on Asian Buddhism. Pursuing her Master's degree in 1986, she elected to experience life in a Buddhist convent in Nepal. Accommodating a friend, on arriving in Kathmandu she delivered a piece of photo equipment to a Thomas Kelly. This meeting sealed her fate. She forsook the convent and pooled her considerable talents with Thomas's. This relationship led to their marriage and their subsequent adventurous life in Asia.

While she was in Nepal, Anita Roddick met Thomas and Carroll. Impressed with their knowledge of Nepalese customs and agriculture, she offered them jobs with the company. They both accepted the intriguing challenge and temporarily moved to England, where they worked at The Body Shop's main office in Houghton, south of London.

Carroll was studying the social anthropology of indigenous people who used the natural herbs the company was collecting. Thomas's assignment was photographing and documenting the lives of the tribes and their use of the plants. He was constantly on assignment in rainforests in Columbia, Brazil, and elsewhere. When he finished with the photography, he returned to England to organize his photos and assemble them into a documentary. Travel for both Thomas and Caroll was frequent and included trips to the United States and Santa Fe.

Anita Roddick was the adventurous product-discoverer for The Body Shop. It was Anita who sent Thomas on his exotic assignments. She liked Thomas and listened when he suggested she investigate the culture of the Native Americans whose pueblos dotted the region around Santa Fe. Intrigued by the idea, Anita came to Santa Fe. She looked to Jeanne and me to show her around and give her suggestions on where to look for products she might be able to use.

Coincidentally, Pamela arrived in Santa Fe from San Francisco at the same time Anita arrived. Pamela, with her new MBA, impressed Anita. Intrigued by the Indians' use of blue corn, Anita commissioned Pamela to investigate the blue corn strain for possible development into a facial powder. Pamela was delighted, accepted the assignment, and worked on the project for a few months. She became friendly with people at the several pueblos growing blue corn and learned from the farmers the uses and characteristics of blue corn meal.

While she was working with the pueblos, Pamela stayed with us in Santa Fe, although she made frequent trips to San Francisco to see her fiancé, Stephen Witt. Stephen was an interesting young man. He graduated from Stanford with a degree in Social Anthropology. He authored several papers about his research in Africa. When Pam met him in San Francisco, he was engaged with this writing.

When she completed her investigation, Pamela, accompanied by Steve, reported her findings to The Body Shop management in England. They were sufficiently impressed with her work to engage her, and later Steve, as consultants of the company. They later became full employees.

Our artist son, Robert, not to be outdone, also established a relationship with The Body Shop and the Roddicks. He made the connection through Celia, a close artist friend of his who had been commissioned to paint a mural in The Body Shop employee restaurant. Celia moved to Houghton and worked on the painting for several months.

Robert visited Celia, as well as Thomas and Carroll, in England, where he, too, became acquainted with Gordon and Anita. A close relationship developed. Although he wasn't employed by the Roddicks, Robert was their guest in England and their host when business brought the Roddicks to New York. Thus a relationship between almost all of the Kelly family and The Body Shop began. (It continued for years to come.)

While the children's careers were unfolding, my relationship with the insurance agency continued. I was available on an on-call basis. When Fred Tiberi needed help with a renewal or the development of a relationship with a potential new client, I assisted. I was, however, becoming more and more an observer rather than an actor. After April, I was no longer paid, but through Fred's generosity, I kept my office space.

Cow Creek continued to be my off-stage passion. Almost every weekend, either alone or with tolerant Jeanne, I continued manicuring the property. Despite the fact that she had her own interests and often wasn't feeling well, Jeanne continued to be a good sport. I was a difficult husband, constantly over-active and basically self-serving and temperamental. Jeanne played her part well to a point; when she had enough of my occasionally thoughtless antics, she would put me in my place, and relative harmony would be reestablished.

During the year, Jeanne suffered several additional, major health difficulties. In June urinary and kidney problems developed. Jeanne tolerated the discomfort, was treated for the condition and resumed her usual routine. In August Jeanne's remaining breast developed symptoms of cancer. Following examinations and biopsies, the diagnosis was, indeed, cancer. She elected breast removal. In September the operation took place and two days later she returned home.

Amazingly Jeanne quickly returned to her active life and immersed herself in preparations for Pamela's nuptials. Pamela and Stephen's marriage took place at Cristo Rey Church on September 17, sanctified by our faithful Father Russo. A festive reception at our home on Palace Avenue followed.

In between bouts of ill health, Jeanne was a welcome and enthusiastic companion with me on two European expeditions. From April 19 through May 10, we toured Russia, under the guidance and auspices of the New York Metropolitan Museum of Modern Art. The reputation of this museum opened the long-shut doors of the principal art museums in the Ukraine and Russia.

Along with Tom and June Catron, John and Mary Dillon, and Kay Harvey and Michael von Helm, we flew to Paris and then, via Russian airlines, to Kiev and Odessa. The newly democratized Ukrainian nation was holding elections. Thus it was a politically active time to be visiting. In Kiev we visited the Puskin museum. In Odessa's harbor, the Ukrainian navy, along with a portion of the southern Russia fleet, rode at anchor.

We spent several days in Tbilisi, Georgia, Stalin's birthplace. Jeanne and I attended services at the Sion Cathedral, the seat of the Georgian Orthodox primate. This ancient church dates back to the fourth century. We toured the Georgian Museum of Anthropology with its amazing collection of artifacts, purportedly dating from 10,000 B.C.E.

We flew to Moscow for an unforgettable tour of the Kremlin complex, a seventy-acre, walled fort and the seat of the Soviet government. I took my morning jog in Red Square, stopping in respect before the guarded tomb of Lenin each time I lapped the square.

After helping the Russians celebrate May Day, we took a train north, visiting Suzdal, the Russian Williamsburg. Then it was on to Novgorod, which was occupied by the German army from August 1941 until January 1944. During that time, the population of Novgorod dropped from 60,000 to 2,300. I think our visit here was prompted by the Russian attempt to impress Western visitors with their own Holocaust at the hands of the German army.

Our tour ended at Leningrad-St. Petersburg, founded by Peter the Great to be Russia's "Window to the West." We toured the Peter and Paul Cathedral, the fabulous Winter Palace/Hermitage complex, the lavish home of Catherine the Great. In recognition of MOMA, the Russian Art Museum, which was under reconstruction at the time, was specially opened for our group.

Filled with a surfeit of Russian food, art and archeology, we left St. Petersburg on May 8 and flew via Pan Am to Frankfurt and on to New York City. After a night's rest, we arrived in Albuquerque on May 9, to be warmly greeted by Pamela and Susan.

Later in the year, shortly after Jeanne's second breast removal operation and Pamela's wedding, I suggested we take a walking tour in Great Britain with an organization called the Wayfarers. To my amazement and admiration, Jeanne agreed and we departed on October 11.

We had been involved with hiking clubs for many years. In the early fifties, the Pajarito Hiking Club was formed, and ten friends and I, all World War II veterans and newlyweds, joined the club's monthly hikes. These day hikes usually ended with our wives or girlfriends meeting us for a tailgate party. The festivities, which always included plenty of rich food, erased the health benefit of the hikes. Regardless, we always had a good time.

As age advanced and families grew, the hikes lessened. By the middle 1970s, regular hikes ceased. In my case, family outings took the place of the Pajarito jaunts. After the departure of the children, the regularity of hikes was reduced even more, although Jeanne and I enjoyed outings and took hikes when camping at our place on Cow Creek. With Jeanne's health problems interfering with our private hikes, I joined the Santa Fe Chili and Marching Society, a group organized by ex-British Brigadier Jack Masters. Designated leaders led highly disciplined, weekly hikes.

Influenced by my experience with the Marching Society, I was attracted to an article describing the "comfortable" walking sojourns of the Wayfarers. I found out this organization was soon to have a weeklong hike in Southern England. I was very attracted by the promise of taking day hikes and spending nights in colorful and comfortable inns along the way. It seemed the routine would suit Jeanne, and an added incentive was the fact that we would be hiking in Devon, where Thomas and Carroll were living while employed by The Body Shop.

We arrived in London on October 12. On our free day before joining the Wayfarers group, we took a cruise down the Thames River to the Barriers, a dam-like structure that prevents the stormy North Sea from flooding up the Thames to London.

We joined our group of ten led by two charming female guides at Claremont on the edge of Dartmoor, a spooky wilderness area in southern Devon. Once off the moor, we trekked along the Channel coast in surprisingly balmy weather. We even spotted an occasional palm tree warmed by the nearby Gulf Stream. We took midday breaks to enjoy good lunches in lively pubs along the route. At night we were guests in small, quaint, yet comfortable inns.

Jeanne hiked only the morning, riding ahead in a van each day to the next designated inn. She thrived and was charmed by the experience. I felt pleased with my decision to attempt the trip so soon after her operation. Our trek ended at Dartmouth, where the Royal Naval Academy was located.

From there, we began the non-hiking part of our tour by taking the train north to Bath where we spent several days touring the city and surrounding countryside. I felt the Bath Cathedral and nearby Wells Cathedral were more beautiful and warmly appealing than the famous but cold gothic churches of continental Europe.

We returned to London on October 23 and spent the next three days exploring that fabled city. Our niece Rachel was enrolled in the Shakespeare drama school. She was staying with her father's friend Steven Norris, a member of Parliament. Norris gave us a full day touring the British Parliament complex, personally guiding us through the House of Commons, the House of Lords, Westminster Hall, and the little known St. Stephen's Chapel. It was a remarkable day made even more memorable by Norris, who at the time was one of Margaret Thatcher's "bright young men."

We spent the final days of our trip with Thomas and Carroll in Houghton. Here we visited The Body Shop, saw Celia's mural in the company cafeteria, and enjoyed being reacquainted with Gordon and Anita Roddick, the company's owners. At the time, The Body Shop enjoyed sales in excess of $100 million and had numerous retail outlets in key cities throughout the world. Since Thomas and Carroll, and now Pamela, were valued employees, we were considered friends of the Roddicks. We became enthusiastic supporters for their continued success.

When we arrived back home in November, the impending U.S. invasion of Iraq was the topic of the day. Our usual day-to-day patterns of activity fell into place. Jeanne seemed to be feeling very well—well enough to ski. I was pleased but still kept my fingers crossed.

1991: Birthdays and Milestones

As the new year arrived, Saddam Hussein's invasion of Kuwait absorbed the country's attention. On February 22 General Schwarzkopf ordered the restive U.S. Army to attack, and the devastating offense began. On March 3 the Iraqi military command agreed to a ceasefire, putting an end to combat. Ours was a magnificently awesome show of military power. Americans were proud, happy, and relieved, despite feeling disappointed that the offensive stopped and did not overrun and occupy the whole of Iraq. I, too, glowed in the general aura of good feeling.

I celebrated my seventieth birthday at a large party where my talented sister-in-law Susan H. Kelly orchestrated a musical rendition of "The Ballad of Bud Kelly." This gala occasion on March 22 was followed by a family reunion and holiday at our favorite place—the La Jolla Beach and Tennis Club.

May 20 was the fiftieth anniversary celebration of Jeanne's graduation from Pine Manor Junior College, near Boston. I was never impressed with the academic reputation of the school, but I was impressed with the pulchritudinous bevy of females making up the student body. Jeanne elected to attend the reunion and I tagged along, as she had done on my many Harvard reunions.

An added inducement to attending the reunion was an auto tour of the Canadian Maritime Provinces with Malcolm and Rosie Stewart. Malcolm had married Rosie after his first wife Mary Helen's death from cancer. We left directly from Boston and spent two weeks driving through New Brunswick, Nova Scotia, Cape Breton, and Prince Edwards Island, then down the St. Lawrence coastline to Quebec and Montreal. In Quebec, we stayed at the fabulous Chateau Frontenac. My rusty French helped a little in Quebec. I had a slight *entré* in that city, as it was the home of the Vingt-Deux—one of the old Canadian army regiments I had known in Korea. We spent the last night of the tour in Montreal attending *The Phantom of the Opera.*

From Canada we returned to Santa Fe and were greeted with the happy news that Thomas and Carroll were to be married in Kathmandu, Nepal, on June 17. Their relationship had been close for several years and now it was time to sanctify the union. We felt it was imperative someone represent the family, but the cost of venturing to Nepal as a family was too heavy a bite off the family exchequer. We delegated Robert, twin brother of the groom-to-be, to go as the family's representative.

It was a wedding ceremony long remembered in Kathmandu, a Newar Buddhist ritual with a female goddess, known as a "Devi" orchestrating. It included a fire ritual and celebrated the natural elements as witness to the union. Following the several hour ceremony, lubricated by copious toasts with *raksi* (a strong Nepali drink), the wedding party mounted decorated elephants and completed the ceremony with a triumphant procession around the center of the city. Robert represented the family so well that, on his return to New York, he took a recuperative leave of absence from his easel.

Not to be outdone, and to legitimize the marital union in her own eyes, Jeanne insisted on a quasi-Christian wedding celebration in Santa Fe. She arranged for this second ceremony and celebration when the newlyweds visited us. Carroll's family flew in from New Jersey to host a rehearsal dinner at the Bishop's Lodge, and the next day there was a mixed Buddhist-Christian procession at our country property on Tano road. Drums, prayer flags, and chanting accompanied Thomas and Carroll's walk to a piñon tree arbor, where Franciscan Father Ricardo Russo cemented the union in the conventional Christian manner. Afterward, the action moved to a large tent with a dance-floor where a wild jazz band swung into action. There were no elephants, but the Tano Road acreage echoed with the celebrating din as the marriage was indeed tied doubly, with both Buddhist and Christian knots.

Unfortunately, Jeanne's health problems continued throughout the year. Her strong and positive attitude and cheerful nature eased our anxiety about these lapses, but shortly after the Santa Fe wedding party, Jeanne became ill with gall bladder inflammation and gallstones. She underwent surgery on September 24 and Thomas and Carroll departed the next day for Kathmandu. Jeanne recovered rapidly and soon returned to her active life. I took pretty much for granted she would continue to be my cheerful companion, accompanying me in our various pursuits.

In late November we spent ten days in San Miguel de Allende, Mexico. My cousin Sol Gross owned a condominium there and graciously offered it to us. We flew to Mexico City and spent two days touring that fabulous but polluted city and then on to San Miguel, which is a haven for transplanted Americans. The gringo colony at the time was ruled by Martha Campbell. She published the news piece *Juarde,* which is a play on a Mexican pronunciation of "Who Are They?" In order to be accepted in social circles in town, one had to have an audience with Martha. Fortunately, we knew her through our friend Rosie Stewart, who was a relative of Martha. Martha greeted us warmly and invited us for Thanksgiving dinner.

While in San Miguel, we visited La Soledad (Monasterio de Nuestra Señora de la Soledad), the hermitage of Father Aelred Wall, who had established the Monastery of Christ in the Desert at Abiquiú. La Soledad was also the new home of Sister Mary Joaquin, the long-time director of St. Vincent Hospital in Santa Fe. Both of these saintly people had retired to San Miguel, but their karma was such that their new locale became a magnet of spirituality focused on the Benedictine monastery and the Catholic medical center directed by Sister Mary Joaquin.

Unfortunately, we were uncomfortable in San Miguel. The modest gas heater did not dispel the chill of the unseasonably cold weather. We had to dress warmly even when we were in the apartment. Getting around required us to walk up and down hilly cobblestone streets, which bothered Jeanne. This concerned me, since she had always been a good hiker. However, despite the discomforts, we enjoyed the experience.

Jeanne seemed to be feeling fine when we returned to Santa Fe. She enrolled in a ski clinic at Taos Ski Valley and spent four days on the area's challenging runs. The Christmas season arrived, along with the children, except for Thomas and Carroll. We ended the year at a New Year's Eve party hosted by John and Sis Dillon at their home in Tesuque.

1992: The Battle

Jeanne's declining health was the focus of our attention throughout 1992. The medical drama was akin to a battle. Initially, there were minor symptoms and skirmishes, but flare-ups soon brought us into intense engagement with the cancer, demanding counter maneuvers—probing and searching and treatments with radiation, chemotherapy, and homeopathy. Temporary lulls in the battle were followed by desperate attempts to avoid being overrun. Then there was the recognition of ultimate defeat with a calming-down before enduring the final, overwhelming enemy assault.

The chronology is painful to recall. During January Jeanne felt poorly but maintained a positive attitude and continued to lead an active, normal life. In February she had a hernia and went through the necessary operation to correct the problem. Concurrently, she developed persistent pain in her chest and ribs. Dr. Riley, her general physician, initiated a series of dye tests and X-rays in March. The test results were unfavorable, revealing hot spots in her lungs, ribs and skull. Dr. Greenfield, her oncologist, declared the condition treatable but not curable. Simply put, her cancer had reappeared, and once the diagnosis was confirmed, we conveyed the sad and sobering news to the children. Treatment began with radiation and chemotherapy.

Seeking for the best possible treatment, we interviewed Dr. Jeffrey Neidhart at the University of New Mexico Medical School. His special treatment regime called for five weeks of intense chemotherapy repeated three times over a four month period. Getting the treatment would have meant moving to Albuquerque. The oncologist Dr. Greenfield felt this approach was too severe.

We sought out another alternative, contacting Dr. Jones, a specialist with Baylor University Medical School in Dallas. He recommended hormonal treatment, rather than chemotherapy. His goal was palliative, intended to reduce discomfort and avoid the effect of the harsh chemotherapy treatment. We felt we were at an impasse, as the doctors had different approaches and theories. Jeanne felt we should leave it up to them to confer and come to a resolution. In the meantime, the chemotherapy continued. When the physicians finally conferred, Dr. Greenfield was dubious about the hormone treatment and recommended a continuation of chemotherapy.

Exhausted and frustrated, in late May we enjoyed a few days of relief in San Francisco visiting Pamela.

Upon returning to Santa Fe, Jeanne's attempt at leading a normal life was interrupted again with severe neck pains. Dr. Feldman, a neurological surgeon—and a warm and compassionate individual—recommended a neck brace for Jeanne, to protect the vertebrae and prevent them from fracturing. Examination revealed new hot spots there, which meant more intensive radiation treatment. Dr. Greenfield also initiated a new chemo approach requiring the implantation of a pump that would infuse drugs for five-day intervals over a twenty-eight-day cycle, to be repeated if results were favorable.

Poor, brave Jeanne suffered without complaining. However, the continual out-breaks of cancerous growths in her body were desperately frustrating for all, especially for our son Thomas. His long and positive experience with Nepalese medicine made him critical of the Western methods used by Jeanne's doctors. His opposition became aggressive and, to some extent, annoying to Jeanne and me.

In the hope that Thomas's homeopathic approach might help—and to assuage his frustration—we agreed to interview his friend, Dr. Yvonne Fisher, a homeopathic physician practicing in Albuquerque. Yvonne was of British and perhaps of Arab extraction. She was attractive, articulate and persuasive. She claimed as patients members of the British Royal Family and the Shah of Oman. Her technique involved homeopathy and acupuncture. Our visit with her lasted all afternoon and included a thorough review of Jeanne's medical and personal history. We departed impressed and somewhat hopeful her approach might be worth trying, with the understanding we would confer with Dr. Greenfield.

Dr. Greenfield felt Dr. Fisher's treatment might, indeed, be helpful and agreed to its implementation, along with his chemotherapy program. Thomas was somewhat mollified. He picked up a large number of small pills and instructions from Dr. Fisher, and the additional treatment commenced. Still, by September Jeanne's condition was worsening. We moved a hospital bed to her second-floor bedroom and engaged Lou Griego as the evening cook. Hospice became involved. A wheelchair ramp was installed from the bedroom down to the living room. During all this time, practically victimized by the rivalry between the treating doctors, Jeanne continued to use the computer, play bridge and have lunch with friends. On occasion she accompanied me to

Cow Creek or to visit friends in Albuquerque or elsewhere. Jeanne's positive attitude buoyed the spirit of her friends and many volunteer caregivers. Although I was the main caregiver, I continued with my routine of playing squash, having lunch with friends, attending business meetings, and other activities.

On my birthday in March, Jeanne and I spent the day at the Red Canyon Ranch, southwest of Clines Corners. We were the guests of Candido Alarid and his family. I had not seen Candy since 1946, when he was one of our favorite cowboys on the Gross Kelly Ranch at Dilia. I gave him his favorite horse, Blue Heaven, when the ranch was sold. Candido, now in his early 70s, had been the working foreman of the Red Canyon Ranch for many years. It was a special birthday treat for me.

In June Jeanne and I spent four days with friends at the Garden of the Gods in Colorado Springs. We enjoyed golf and tennis and visiting with our old friends the Stewarts, the Wilhelms, and the Malos. Later that summer, I joined Phil Schultz, Wooly Henry, and David Hughes on a four-day pack trip in the Weminuche Wilderness of Colorado. Tom Van Soelen was the outfitter, assisted by his father, Don Van Soelen, our longtime friend. I was a casualty on this trip. I twisted my right knee and tore the meniscus lining. This ultimately required surgical repair. Thus, for a time, I joined Jeanne in the homecare department.

Remarkably, Jeanne seemed somewhat better in November. When we celebrated our thirty-ninth wedding anniversary on November 28, she felt well enough to lunch at Café Escalera and to dine that evening at El Nido in Tesuque. But on December 18 Jeanne had a serious relapse. Her pain was severe and she may have been overdosed with morphine. In any case, she went into a deep coma for several hours. She came out of it, but it was a telling lapse. Dr. Greenfield reluctantly advised Jeanne she had only a few days or weeks more before the end.

Jeanne seemed to rebound for the Christmas holidays, when all of the children came home. As was our custom, we put out *luminarias* and *farolitos* on Christmas Eve. I took Jeanne to the children's Mass at Cristo Rey. On Christmas Day, presents and children surrounded Jeanne in her wheelchair in the living room. That evening, we enjoyed a roast beef dinner with our family and close friends. It was a festive and special dinner, especially enjoyed by Jeanne, encircled by her devoted family.

On New Year's Eve the younger set went out while Jeanne and I watched a movie on our new VCR. We toasted each other with champagne, I gave her the last evening pill at midnight, and so ended another year.

1993: Adjusting

Jeanne was struggling to stay alive. She wanted to go on living, but the sentence was final. Jeanne died on Thursday, January 21, 1993, at fifteen minutes after midnight. All her family and the wonderful hospice nurse surrounded her bed. Death had taken over her worn and tired body, but her spirit had been released, and relief seemed suddenly to permeate the room.

Thomas and Carroll read from Buddhist book *The Tibetan Book of Living and Dying* and performed several rituals in which we prayed for the smooth transition of Jeanne's soul to its next destination. It was as though a beautiful butterfly suddenly appeared fluttering above all of us, searching for the way out of confinement. With small bells, feathers and incense, Thomas and Carroll chanted, invoking aid and comfort to Jeanne's soul and at the same time giving the grievers a sense of release. This ceremony broke the bubble of sadness and inaction and initiated a spontaneous, positive response in all the children.

Following an extremely short night, the children orchestrated a *velación* (vigil) for their mother. Thomas and Carroll wrote an obituary. Robert secured the Guadalupe chapel for services while Susan and Pamela arranged the centerpieces in which we would place the cremation urn.

By mid-afternoon the visitation began. Some two hundred friends paid tribute to Jeanne and visited with the family. It was a gracious and heartfelt recognition of the widespread affection felt for her.

On Saturday, January 23, Father Ricardo Russo celebrated the funeral Mass at Cristo Rey. Guitar music evoked tears as some of Jeanne's favorite pieces were strummed. It was a moving ceremony, followed by a gathering at our house on Palace Avenue. The curtain came down, the drama was over, and for me suddenly there was silence and a poignant aloneness, yet relief.

The previous six months had been completely absorbed by Jeanne's illness. Putting up with doctors, examinations, pain, pills, nurses, treatments, and a constant stream of visitors and caregivers was exhausting, certainly for Jeanne, but also for me. I was essential to the caregiving process but often felt redundant in my own house and home.

My initial reaction to a solo existence was curious. I greeted my new life with a sense of relief and a strong desire for solitude. I was anxious for the children to reconnect with their individual lives. I relished the idea of having my own way. I was tired of being a host and provider. I felt somewhat guilty about having these self-centered thoughts and desires, but that was my reaction.

Shortly after the funeral, we had a family meeting. Jeanne had left several bequests, with which we complied. We divided her jewelry and personal items among the children, friends, and Rose, our longtime housekeeper. Surprisingly, I spoke of my desire to have the children consolidate and remove their residual childhood belongings, which had remained in their rooms, closets and elsewhere in the house. I wanted the home to be mine—and the children to be visitors and not part-time occupiers. They all had their own homes and they all realized this request was fair, but it did take them by surprise and made them realize things had changed.

Notwithstanding my house-clearing ultimatum, Jeanne's illness and death brought my relationship with the children much closer. During the critical stages of Jeanne's illness, I was impressed with their warmth and obvious affection for her. Instead of being a devoted but somewhat critical parent, I became more of an intimate adult friend. I sought out their opinions and feelings. All this altered the relationship equation. I felt I could let go of the role of controlling parent.

My adjustment to life as a widower was akin to a relapse into bachelorhood. I didn't slip into being a lonely and inexperienced, mate-less housekeeper. Jeanne and I had maintained a relationship of mutual dependence, but at the same time we kept up independent activities. For her part, Jeanne was a very competent, loving, and practical person. She ran the domestic side of our marriage smoothly and comfortably. She paid the bills, managed the house and gardens, supervised the children's activities and enjoyed entertaining. A good athlete, she also liked sports, playing golf and tennis, swimming and skiing. Jeanne had her tennis and skiing buddies, she loved to play bridge and she kept busy with study clubs and church affairs, as well as museum functions.

My side of the relationship was more male oriented and tied up with work obligations. The insurance business was quite demanding. I had many clients and was constantly busy maintaining close business relationships and obtaining new accounts. Business lunches were almost a daily occurrence. I had to constantly massage relationships with the insurance companies that sought the agency's business. All this was essential for the agency's success.

Public relations and pro bono involvement in civic affairs also were essential and dutiful parts of my business activities. These demanded frequent meetings that consumed considerable time. Then there was the Cow Creek property, where I spent many weekends in the summer, and all the sports I enjoyed with my male friends.

Thus, despite Jeanne's death, I had plenty to keep me busy. The main difference, and initially a critical one, was the absence of companionship and the affectionate communication that came with our happy marriage. Jeanne's long and debilitating illness had cut into this aspect of our lives, however, so by the time of her death, that facet of our relationship had been much diminished.

The children felt Jeanne's loss deeply, especially Susan, the only one of our children living in Santa Fe. She was single, so I went with her to regular bereavement meetings sponsored by Hospice. I did this more out of a desire to help her than to alleviate my feelings, but we both found these tearful, heartrending gatherings to be quite therapeutic. They helped Susan with her grief and encouraged in me an inner awakening.

Jeanne's devoted friends responded sympathetically to my new condition. The number of invitations I received to dinner and other activities increased soon after her death. My efforts to teach myself to cook were hampered by these much appreciated invitations.

Among my friends was another recent widower, Burch Ault. Seeing he needed consoling, in March I invited him to join me for a visit to La Jolla. It was a refreshing change for both of us. After his return to Santa Fe, I lingered in San Francisco, where Susan and Pamela joined me. When I returned home, I continued to acclimatize to bachelorhood. Weekly hikes with the Santa Fe Chili and Marching Society warmed my friendship with Wooly Henry and John Baxter. In May we three decided to hike the Grand Canyon. We drove to the South Rim of the Canyon, left our car there, and took a long shuttle drive to the North Rim, some two thousand feet higher than the South Rim. We had previously secured beds and meals at the Phantom Ranch at the Canyon bottom. The hike from the North Rim to the ranch at the Colorado River was an exciting all-day venture.

We were assigned bunks in the men's bunkhouse at Phantom Ranch. I had the top slot, Baxter the middle, and Wooly the bottom. After dinner, exhausted, we settled in to rest, but the bunkhouse lights, left on for occupants who were yet to retire, glared down on us. Wooly fell asleep, but Baxter and I lay awake. We noticed a long snake slithering on the floor, approaching

Wooly, the Sleeping Beauty. Seeing the snake was a harmless, non-venomous bull snake, we let it proceed across the floor. Only when it was at Wooly's beside did we alert the snoring Wooly. Wide-eyed, he stared directly into the eyes of the snake, obviously scared to death. We, the observers, almost died laughing, recalling that Wooly somehow always attracted snakes. The snake was tossed out of the bunkhouse and we spent the rest of the night in peace.

We climbed out of the Canyon via the South Rim trail the next day and returned to Santa Fe. Shortly later, I attended my fiftieth reunion at Harvard. The ranks had thinned considerably. It was a sentimental return, enlivened by a younger group that included my son Robert, there for his twenty-fifth reunion.

Later that summer, somewhat restless and being completely free and independent, I decided to attend the Harvard seminar at the University of Cambridge in England. The seminar focused on Elizabethan England and John Harvard's student days at Cambridge in the early seventeenth century. The thirty attendees were housed at Emmanuel College, Harvard's college when he attended the university. It was an unforgettable experience, with daily lectures, visits to the surrounding historic sights, and exposure to the beauty, tradition, and history of Cambridge University.

The children came home for the Christmas holidays. Their presence enlivened the house and returned some of its former warmth. We decorated it with ornaments Jeanne had liked. Christmas dinner was enjoyed potluck style with guests Booker and his family, the Grabs and Parkers, Barbara Hooton, Marigay Graña, Mary Ellis Alley, and Burch Ault.

Jeanne, absent in person, was with us in spirit.

1994-2000

NEW PASTURES

1994: The Age of Adult Puberty

In 1994, a year after Jeanne's death, I again became interested in the opposite sex. It was like adult puberty. While I was married, I had maintained friendly relations with women. I enjoyed their company and liked some more than others, but I only associated with them in activities that included Jeanne. She was popular with my male friends, as I was, to some extent, with their wives and lady friends. I never strayed far from the path of devoted fidelity during our thirty-nine years of marriage.

After Jeanne's death, I was often the single male invited to gatherings and parties to balance the male-female ratio. If I was invited out to dinner or to escort a lady to a function—and if I found the woman attractive—I felt obliged to reciprocate. As time progressed, I became friendly and comfortable with several widows and single women. They did not object if I asked them on a date or to join me in some activity.

Over time, though, these social relationships with women became more complicated and at times of concern to me. Women seemed not to appreciate female competition. I was careful to avoid embarrassing my women friends by running into other female friends or by offering conflicting invitations. At the same time, I was not interested in an "affair" or going strictly "steady." Being an insurance professional, minimizing risk was deeply engrained in my psyche. I welcomed female attraction but resisted commitment.

My first relationship started up in late February when I was in Florida visiting my close friends, the Downers. At a dinner party, I was paired with an attractive widow named Aileen Gavin, who lived in Chicago and wintered in Florida. Her friends and my hosts felt we were ideally matched. I enjoyed her company and saw her several times during my stay. The Downer invitation was initially for a week, but Aileen suggested I extend the stay as her guest. It was tempting, but I resisted and returned to safe ground in Santa Fe.

I next became involved with Joan Curwin, the widow of a Harvard classmate. I had met her at reunions and had seen her in Boston while en route to Cambridge, England, the previous year. We corresponded occasionally. In a couple of letters to her, I casually mentioned that in late March and early April I was taking a cruise around New Zealand with an extension to Australia. When I arrived in Auckland, New Zealand, to board the cruise ship, I found she, too, was a passenger. This was a complete surprise, and not an unpleasant one, but I felt it was devious on her part to show up like that. I now was locked on to a companion for the whole journey, whether I liked it or not.

On the cruise, Joan and I were definitely a pair, sitting together on buses and at dinner tables for the circumnavigation of both the north and south islands. Then came the Australian extension. We toured Sydney, then flew to Alice Springs and to Ayers Rock. Our relations were affectionate but not heated. By the time the trip ended, I longed for fresh, single isolation. That may also have been the case with Joan. We parted amiably at Sydney and I returned to neutral but barren ground in Santa Fe.

I recommenced the chessboard relationships with my lady friends in Santa Fe. I became involved in my close friend Joyce Merchant Peters's successful battle with breast cancer, an experience that reminded me of my caregiving days with Jeanne. Still, I resisted becoming serious with her or any one else, although several of my single male friends succumbed and remarried. Burch Ault was among these. A reluctant bachelor, Burch was released from his unwanted state by marrying the widowed Florence Monks—the start of a happy relationship that was to last many years.

My daughter Susan became engaged to Christy Stanley in June, with a wedding planned for July 30. I was responsible for Susan's wedding preparations. She wanted the whole enchilada. Fortunately, friends of Jeanne and Susan assisted me, and in any case my most important duty was to give Susan away and pay the wedding expenses. Ladies like weddings, and this one was planned to a T.

The wedding ceremony took place in Susan's grandparents' old home, adjacent to mine, and the reception was tented on the lawn at my house. Everything went smoothly, except that the plumbing, over-taxed by some 250 guests, succumbed, and I, as the father of the bride, struggled in vain to remedy the situation with a plunger. Susan's brother Robert, the artist, interrupted his dancing and fashioned a No Toilet sign at the main entrance to the house. Luckily, the sanitary ban did not curb the frivolity.

August was a tortuous month. Wedding guests lingered. Robert was in from in New York and busy preparing an art show for the Linda Durham Gallery in Santa Fe. Thomas, also here for the wedding, used my house as his Asiatic command post. He was constantly either on the telephone or computer managing his one-man, worldwide operation while at the same time greeting a stream of visitors from the Orient.

I had maintained an office hideaway at the Kelly Agency for years after I sold the business. However, to further complicate my life, the agency was suffering and was being sold. I lost my sanctuary and had to move my files and workplace to my beehive-busy home.

Thankfully, the pressure eased by September. Thomas returned to Nepal, Robert to New York, and Pamela to San Francisco. Susan was busy keeping her new husband happy. I continued my weekly hikes, attended bank meetings, and usually worked on weekends at the Cow Creek rancho. Socially, I continued to be quite active, enjoying my retired bachelor status.

I spent Thanksgiving day with Susan and Christy and the senior Stanleys in Socorro, with a traditional turkey dinner at the Valverde Hotel. From there, we drove to the Bosque del Apache wildlife refuge to see the migrating waterfowl. The next day we drove to Hillsboro, an old mining town, and then to scenic Marlboro Canyon, which knifes through the rugged desert mountains. The primitive road finally became the Alamosa Creek bed, which led through a passage barely as wide as a car into a wide, empty valley. Several meandering warm springs watered the valley, a refuge that in times past was the winter campground for Geronimo's Apaches. With the subjugation of the Apaches, it became the Warm Springs Apache Reservation, with a U.S. Army fort named Fort Harmony—hardly an appropriate name for such a remote place and frustrating duty assignment. This place was a memorable discovery for me and brought with it the haunting realization I had much to be grateful for on this Thanksgiving weekend. Fittingly, it was also the forty-first anniversary of Jeanne's and my wedding and the happy day that Susan announced her pregnancy.

The holiday season brought the children to Santa Fe for the customary annual family gathering and Christmas festivities. I was the lone parent, somewhat selfishly protected by my children. I think they preferred me single, rather than paired up with a new woman friend. I concurred. It had been, however, an adventurous year probing into female territory.

1995: Diary Reflections

My first diary entry for 1995 states, "The year bodes good and bad—Susan pregnant, Pamela divorcing, Robert working to scale up his painting, Thomas and Carroll involved in writing, photographing and cultural assignments in Nepal. I am not sure of my direction, except for the fast approaching trip to Australia."

As I progressed through my seventy-fourth year, I was quite healthy, active and alert. I was playing squash regularly, hiking almost every week, tending to house and garden, monitoring my finances and serving as a director on the boards of three organizations. In addition, I was quite active in my social life, continuing to pursue several lady friends. Things were going quite well, but nevertheless I fretted about my health.

My immediate concern was the logistics of a trip to Australia with Joyce Merchant Peters. Her son Pike was to be married to an Australia lady in Melbourne. Joyce asked me to accompany her to the festivities. Instead of being thrilled, I was deeply apprehensive about the threat of intimacy. I wanted to take the trip but at the same time I wanted my own space, including bedding. I also wanted my freedom to avoid excess involvement in the affairs of the wedding families. I conjured up all sorts of embarrassing situations that might arise and became quite worked up over these phantom possibilities. Fortunately, Joyce had an almost psychic ability to sense my fears and addressed them as she planned the venture.

On January 23 we flew from Los Angeles to Sydney, covering the 12,000 kilometers in fourteen hours. After an hour-long layover there, we took off for Melbourne, where we met the groom, Pike, and the bride, Marianne Stillwell. Since the wedding was not scheduled for several days, we took the opportunity to visit Tasmania. We flew to Hobart, Tasmania, about an hour south from the Australian mainland on the south side of the island. Beyond Hobart, the ocean stretches to Antarctica.

Tasmania suffers from the blight of being the historic site of prisoner incarceration for the British Empire. Prisons were situated on the Tasman Peninsula, which is attached to the main island by a scant thread of land about two hundred yards wide. This narrow strip of land was guarded by vicious dogs that prevented escape from the peninsula. This made it a convenient and secure dumping ground for prisoners. Guilty and not-so-guilty convicts were banished there in the nineteenth century. They were treated brutally, and despite the fact that this inhumane treatment was discontinued by the 1860s, just the memory of it left an enduring impression on us in a strong and depressing way.

Fortunately, the rest of the island held much interest and scenic beauty. We hiked to Wine Glass Bay, a beautiful, isolated sandy beach kept inviolate as part of a nature park. Farming and timbering are the mainstays of the economy. One unfortunate impact of the forest industry is the considerable presence of mashed, muskrat-like Tasmanian Devils, victims of logging trucks.

On returning to Melbourne, we watched some of the exciting matches at the Australian Open tennis tournament. As the wedding day approached, the social momentum increased. It was a strategic time for me to dodge some of this activity. Intent on seeing as much as I could of Australia, I flew to Perth, the capital of Western Australia. The four-hour flight from Melbourne passed over the vast, sparsely inhabited deserts of Australia—roughly the distance from New York to San Francisco. (So large is the Australian continent that all of the United States, minus New England, could fit within it.) Perth's economic focus is Africa, across the Indian Ocean, and Indonesia to the northwest.

In Perth, I was the guest of Peter and Anne Jooste, South Africans from Cape Town who emigrated to Australia. Peter was a prominent lawyer and Anne was deeply involved in the cultural life of Perth. Before moving to Australia, Peter had spent several weeks in Santa Fe on a Rotary Club fellowship. We had become good friends and had kept in contact after his move to Australia. Anne Jooste was my tour guide and showed me much of the Perth area. We spent a day on the beach fronting the Indian Ocean.

Returning to Melbourne, I escorted Joyce to the marriage ceremony and the fabulous wedding reception at the Stillwell estate outside of Melbourne. The Australians are not stuffy. Hot jazz and famous Australian wine and beer lubricated the celebration. It took me a couple of days to recover as I visited museums, botanical gardens, and the incredible beaches along the Indian ocean.

By the end of the wedding, Joyce and I had been together for most of our two weeks in Australia, except for my trip to Perth. At the end of the adventure, both of us needed our space, although we remained good friends. In Santa Fe, I retreated to my bachelor den to savor the relative isolation and to recover from jetlag. I resumed my familiar routine. Despite my caution against becoming seriously involved with women, I went on dates with Marigay, Joan Bokum, and Bobbie Hilliard, taking them out to pay them back for home-cooked dinners they prepared for me.

On February 22 I attended Mass at the Carmelite Convent in honor of Jeanne's seventy-third birthday. I celebrated my birthday on March 22 with Laughlin and René Barker as a guest in their RV. We visited Silver City and the Gila ruins, Cloudcroft and the Mescalero Apache Reservation. In April I began taking Italian lessons along with Leni and John Baxter in anticipation of a trip to Italy that Leni was planning. My sons Robert and Thomas turned thirty-nine on April 16, but I couldn't congratulate them, since Robert was in Costa Rica and Thomas was in Nepal. I attended a sentimental reception at the New Mexico Museum of Art on April 28, the eighty-fifth anniversary of the founding of the Women's Board of the Museum of New Mexico. My mother, my wife, and my daughter Susan had all been members.

In May, Rosa Cruz, our long-time family maid, retired. Her mother, Incarnación Rodriguez, and her aunt Josefita Gorman had preceded her as family retainers. Rose did not approve of social security, so I gave her one thousand dollars in recognition of her long and devoted attention. She deserved much more but was pleased with the remembrance. May 8 was the fiftieth anniversary of the Allied victory in Europe, a day I will never forget.

On May 13 we left on our long anticipated trip to Italy. The genesis of this adventure was a reward Leni Schwartz Baxter received for her excellent decorating of the Van Camp Zecca residence in Galisteo. The reward was a three-week stay at the Van Camp's Castellano Uganan, near Greve in Tuscany. Leni had invited Wolcott and Beth Henry and my daughter Pamela and me as guests. We flew to Pisa via England and to our palatial home near Greve.

The estate was a working olive farm and vineyard. The main residence could accommodate six guests, and an adjoining guesthouse had room for four. Roberto Zecca oversaw the domestic staff of maids, kitchen help, and chef. The farm included an ancient Etruscan road, which led from the Castellano down to Greve. Wolcott, Baxter and I, all members of the walking club in Santa Fe, took this hike almost daily. As a treat, we feasted on the delicious gelato at the road's end in Greve.

Our host Roberto was an excellent tour guide and a gourmet. He also greatly admired Pamela. He introduced us to several quaint and delectable restaurants in the area, which generally were unknown to ordinary tourists. We visited San Gimignano, Perugia, Florence, Sienna, and several monasteries and their accompanying wineries. The experience converted us to lovers of Tuscany.

After the Australian and Italian trips, both taken before the middle of the year, the rest of 1995 was something of an anticlimax, but there were memorable events. Susan gave birth to a beautiful granddaughter named Caroline, my first granddaughter. Susan allowed her brother Robert and me to witness the drama of the delivery. This was a new experience for me. In my fathering time, husbands were generally kept away from the delivery room until after the child and mother were somewhat composed. Now we watched as Christy assisted Susan in labor and then cut the umbilical cord from the newborn child. His attention to Susan in this amazing process vividly demonstrated the responsibility he felt in parenting. Watching it made me believe all fathers should be present at their children's births, although few fathers could outshine Christy.

One of our weekly hikes later in the year, led by John Baxter, was particularly inspiring to me. We walked to the remnants of a Spanish village known as San Lorenzo, located on the Rito del Oso in the high country west of Española. We drove for thirty-nine miles from Española and began the hike where the road reaches the Rito del Oso. Three or four miles down the drainage we came to a meadow with the stream meandering through it and a clump of spike-like apple shoots pointing skyward. These trees are the remnants of an orchard planted by the Spanish pioneers who settled this tiny outpost in the early 1700s. Around 1730, Utes raided and demolished the outpost, leaving only the apple trees as evidence of the former settlement. Seeing these relics vividly brought to mind the perils faced by early colonists.

In the fall Pamela began working with Nancy Meem Wirth, developing a furniture business to sell replicas of door designs created by Nancy's father, John Gaw Meem. This experience gave Pamela the idea of marketing some of the Museum of New Mexico's unique collection. This led to discussions with the principals of the New Mexico Museum Foundation. Commuting from San Francisco, she was in Santa Fe for the Christmas season. Pamela and I hosted a Christmas season party, entertaining thirty-five guests, some of whom were her museum friends. It was a fitting prelude to Christmas and one I welcomed, as Pamela was the only member of my family to be with me on Christmas day.

On New Year's Day, I was a guest at Betty Martin's party. It was snowing, and on returning home, I noted in my diary ... "I ruminate in the quiet. Another year ended. Robert, Thomas and Carroll are in Goa, India; Pamela is skiing at Monarch in Colorado; Susan, Christy and Caroline are in Socorro; Jeanne is in heaven; and Tika (my dog) and I are at 535 Palace Ave."

1996: Excursions and Interludes

The winter routine was broken by a weeklong visit in January to Florida with Joe and Louise Downer, prosperous "snow birds" from Long Island. Joe owned a comfortable cottage at Loblolly on Jupiter Island north of Palm Beach. I had visited there before and was familiar with the lifestyle. Joe liked walking on the isolated beach a short boat ride from the house. I enjoyed it, too. We could reminisce about our lives, deplore the state of the nation, view the ocean birds and dip into the friendly, lapping water, if we felt the urge. Louise enjoyed a different set of interests: golf, tennis, and al fresco club lunches. My level of play in golf and tennis was about equal to hers, so we were often matched together. My visit with Joe and Louise offered a good combination of activities and companionship. A week was just the right length of time to ensure the continuing freshness of our relationship.

In February my Santa Fe friend Joan Bokum was a houseguest in Litchfield Park, Arizona, of her close friends Cathy and Warren Morton from Cheyenne, Wyoming. I had previously met Cathy and Warren in Santa Fe. At Joan's suggestion, I was invited for the weekend of February 15-18. Joan was persuasive, and I accepted.

Litchfield Park, in the desert about fifty miles from Phoenix, is a second-home community offering golf, tennis and bird hunting. Warren was an avid golfer as well as an accomplished bird hunter and dog trainer. It all sounded great, but I had some doubts about the visit. I had struck a diplomatic balance with my several lady friends in Santa Fe, a policy of neutrality with no obvious partiality. I feared spending this weekend with Joan could disrupt that balance. Furthermore, my golf prowess was limited. I only played an occasional game with equal duffers and was no match for Warren's golfing group. As for hunting, I had no experience with retrievers and was, in any case, only a mediocre marksman. That left tennis. I was a fair player and perhaps could win more games than could Cathy or Joan.

After some indecision, I accepted the weekend invitation and had an excellent time, mostly with the ladies. It was a low-key visit and it did not upset the rotational balance of harmonious female relationships in Santa Fe.

On March 20 in Kathmandu, Nepal, Carroll and Thomas became parents of a baby boy they named Liam Dunham Kelly, making me a grandpa twice over.

In early April I visited the Gray Ranch, down in the "boot heel" of New Mexico. I had always been intrigued by the area, a mapmaking anomaly created when the Gadsden Purchase of 1853 created a "heel" of U.S. territory that projected southward. The area shares a remote boundary with New Mexico, Arizona and Mexico. The famed Gray Ranch lies between the Animas and Peloncillo mountains, right up against the border with Mexico.

Joyce Peters, whom I had accompanied to Australia in 1995, was familiar with the region. Her great-grandfather Claiborne Merchant had ranched in the area in 1883, and Drummond Hadley, the new owner of the Gray Ranch, was her friend. I had just previously met Drum and he set up a visit to the Gray Ranch when I expressed a desire to see it.

My daughter Susan, her husband Christy, Joyce and I started out on this adventure on April 3. Christy, an excellent driver, steered his van through Socorro and on to Hatch, Deming, Lordsburg, and into the Animas Valley. At the Gray Ranch, headquarters of the Animas Foundation, a nonprofit whose mission is to maintain the ecological integrity of the area, we met the foundation manager, Bennett Brown. The 325,000-acre Gray Ranch was protected from overgrazing and much of it remained pristine because of Drum and the Foundation's careful management. Additionally, the foundation helped local ranchers, who were prime targets for "ranchette" developers when drought and/or an adverse economy threatened to force the ranchers to sell their land. To prevent this undesirable subdivision of the ranchland, the foundation would "lend" pasture from the Gray Ranch land bank to the troubled ranchers, thus preventing a forced sale.

Bennett described the foundation's mission and gave us general directions to places we could visit on the ranch. The playa valleys between mountain ranges were heavy with grass, watered by the pools of water impounded from mountain runoff. Groves of juniper, oak, and cottonwood trees separated the playa meadows. The tenants of the land were antelope, javelinas, and white-tailed deer. We were the only human trespassers, except for four U.S. government predator-control officers hunting rogue coyotes. When we finished the tour, we reluctantly turned back north, left the ranch and drove to the civilized hamlet of Rodeo, New Mexico, where we spent the night. En route back to Santa Fe, we detoured to Skeleton Canyon, near Rodeo, a haunting, remote canyon where Geronimo finally surrendered to U.S. Army troops under General Nelson Miles in 1886.

Less than a month after the Gray Ranch expedition, I was in Spain with a study group organized by Tom Chávez, director of the Palace of the Governors. We were traveling to Trujillo, a town of nine-thousand people in west-central Spain, an area known as the "nursery of the conquistadors." Trujillo was the birthplace of Francisco Pizarro, the Spanish conqueror of the Inca empire in Peru. Our academic host was the de Salas Foundation and our Spanish language classrooms were in the Museo de La Coria, a former *convento* that overlooked the city.

Most of our group, seeking local color, elected to stay in rental facilities in the vicinity of the Plaza Mayor, but I chose the government-run Parador Nacional de Turismo, more removed from the plaza. Formerly the Convento de San Miguel, it was comfortable, with central heating and a well-run restaurant, even though it lacked the flavor of "normal living" in Trujillo.

The group was divided into sections depending on familiarity with the Spanish language. Our instructors were imports from Santa Fe. Mine was Jim Gavin, a fluent linguist and an able teacher of Spanish. Our classes ran from roughly eight in the morning until noon, leaving the rest of the day free to take organized trips in the surrounding area.

Unfortunately for those who chose to reside near the plaza, the weather often proved to be the enemy. It was cold and damp. Most of the rentals lacked central heating and hot water. As a result, my warm and comfortable Parador room became popular. This was especially true for several of the ladies who sought my room's heated radiators to dry their laundry. One in particular was Sandia Johnston, who became a good friend.

Our group members were compatible. After class, we generally would venture to the Plaza Mayor for lunch and in the evenings would congregate at one of the bar/restaurants frequented by the locals. The city itself was interesting.

Pizarro's palacio was a prominent anchor to the plaza. The Castillo, or fortress, which commanded the local promontory, was safe harbor for Christians during the Moorish occupation of Spain. Hikes outside of the fortress-city were exciting, as the area was dotted with watchtowers formerly manned by lookouts on guard against Moorish raiders. For us, the watchtowers served as observation posts and havens if one of the loco *toros* (bulls) appeared menacing.

Complimenting our language studies were trips to various cities and places of historical interest. We visited Cáceres, Guadalupe, as well as Tordesillas and Salamanca. Of particular interest for us Santa Feans was Guadalupe, where the original basilica of the Virgin of Guadalupe was located. I was intrigued by Tordesillas, where Juana la Loca was placed in safekeeping and where the Pope divided ownership of the New World between Portugal and Spain.

From Salamanca, we took a four-hour bus trip to Madrid and the airport. I returned safely to Santa Fe, exhausted but relishing this exciting exposure to Spain.

Despite what seemed like a constant series of trips, there were major intervals in between them in Santa Fe. Almost each week involved a hike, as I was the president of the Santa Fe Chili and Marching Society and thus had to set the proper example of punctuality. Additionally, the Chili Club met monthly, requiring attendance and some preparation. I usually spent a day each week and sometimes stayed overnight at Cow Creek, working on improving its timber stand. I continued as a bank director and also served on the board of the Kiva Club. I did not lack for interests.

In October the children arrived for what had become an annual family business meeting. This year the subdivision and the allocation of forty acres I owned on Tano Road was the main object of the meeting. The property had been resurveyed and divided into four equal plots. Selection of ownership and protective covenants had to be determined. The meetings achieved their purpose. It was a pleasant reunion and enabled the scattered family to become cohesive again.

My last excursion of the year was a pre-Christmas, long weekend in San Francisco with daughter Pamela. I was the honored guest at a party she gave. Pamela took me to Nordstrom's to re-outfit my worn wardrobe, and in return I helped her finance the purchase of a home in San Francisco.

1997: Resolutions

I usually started a new year with a list of firm resolutions. Over the years, the list suffered greatly from continuous repetition. This year I resolved to be more compassionate, thoughtful, and empathetic with others, including my children and their friends. I also planned to work on the family genealogy, to consolidate family photographs and to construct a cabin at Cow Creek, drill a new well and put the property into a trust. My usual commitments to keeping fit, informed and involved, completed the prospectus.

Erosion of these ambitious goals started almost immediately. On January 15 I departed on a South American tour organized by the Council on International Relations, led by Rose Carlson. My tour companion was Marigay Graña, a close friend from pre-widowhood days with whom I had since established a fairly intimate alliance. She had adopted New Mexico as her second home. She owned a remote cabin/rancho in San Miguel County. She was well educated, familiar with Latin America, and was politically well informed, though extremely liberal—a holdover from her Berkeley, California, hippy days. It was fortunate we were not traveling roommates, as our political differences would have demanded nighttime separation and rehabilitation.

Politically and economically, Chile was enjoying a reprieve from the militaristic discipline of General Pinochet's day. A reasonable democracy was in place, though monitored carefully by the retired but widely admired former President Pinochet.

Our stay in Santiago and vicinity included people-watching and visits to museums, cathedrals and restaurants. We detoured to Isla Negra and visited the beachside home of Chile's hero-poet, Pablo Neruda. There was a revival in interest in his poems at the time of our visit. Chilean wines enjoyed a reputation for excellence, especially because of the products from the Rothschild vineyards and winery, which we visited.

After three days in Santiago, we endured an all-night bus ride south to Puerto Montt, then an airplane flight to Coyhaique and a bus ride to Chacabuco. We entered the Chilean glacial area and boarded the Patagonian Express, a catamaran that delivered us to the isolated but elegant Hotel Termas de Puyuhuapi, a hot spring resort and gateway to the bay containing the San Rafael glacier. This trek took its toll. Marigay suffered from a stomach disorder and fatigue she tolerated only by drinking shots of fiery pisco sours. I suffered, but only from the lingering punch of the piscos.

Following a day of rehabilitation in the warmth of the resort's mineral baths, we again boarded the catamaran to take a serpentine course through ice flows and islands to the vicinity of the San Rafael glacier. During the voyage, I was fortunate to be an observer on the ship's bridge, watching the helmsman guide the ship through the maze of obstacles. Once we were near the glacier, we put on waterproof gear, divided into small groups, and boarded zodiacs for a choppy and freezing trip to view ice blocks calving from the mother glacier. From a distance of some two hundred yards we watched these tremendous chunks of glacial ice slam into the frigid sea. Once we were back on the catamaran, we were rewarded with shots of scotch whiskey laced with ice chips taken from the newly calved icebergs.

We returned to Puerto Montt and then spent the day riding buses and ferryboats through Chilean national parks into Argentina to arrive at the ski resort of San Carlos de Bariloche, with its Swiss/Bavarian style architecture. Here, Marigay and I detoured from the tour to make a trip to Patagonia, where my cousin Adam Canapa and his wife, Kyle Singer, owned and operated La Rinconada, an adventure inn in the lake district. I was anxious to see the ranching country of Patagonia and had made reservations to visit Adam for a few days. He met us in Bariloche and drove us for three and a half hours to reach La Rinconada.

The eight-hundred-acre resort is adjacent to several large lakes that border the snow covered Chilean Andes immediately to the east. On the west is Patagonia, the great undulating plains, and some of the prime sheep and cattle range of Argentina. For three days our hosts entertained us, including an elaborate outdoor party where we feasted on roasted lamb and wine of the region. We also traveled to the nearest village, Cholila, to see a county fair, complete with local horse racing. Watching the gauchos compete, drink and bet was contagious. What we lost in our bets on the races was made up for by the good-humored friendship of the contestants.

Adam and Kyle drove us back to Bariloche and we flew to Buenos Aires, rejoining the group for four days in this sophisticated, sensuous, and politically discontented city. Buenos Aires at the time was recovering from the hangover of the Perón era. President Menem was a reasonably popular president. The peso was tied to the dollar in value. The economy was in a distressfully conservative period of fiscal restraint. Despite this Lenten atmosphere, the city and its people seemed well fed and, generally, to be passively accepting their lot.

One sobering sight was the daily pilgrimage for the *desaparecidos*, led by mothers who gathered at the Plaza de Mayo, seeking solace and some information about their "disappeared" husbands and sons. We paid a visit to the Recoleta Cemetery, a popular destination where Evita Perón is buried. The tearful song of *Don't Cry For Me Argentina* was continually on the radios. Somewhat overloaded with museums and churches, our last night was libidinous, as we were fixated on entrancing tango dancing at El Viejo Almacén.

The tour group, including Marigay, went on to Peru while I, weary of traveling, departed Buenos Aires for a return to Santa Fe. I resumed my routines, although life was anything but static or boring. My children were all leading their own lives. I remained concerned for their well-being, although I was now an interested observer. Susan delivered her son Timothy on January 31, with Christy again practically the midwife. Robert had two successful art shows, one in Sun Valley and the other at Linda Durham's new gallery south of Santa Fe in Galisteo. Pamela, although a new homeowner in San Francisco, was in the midst of negotiations to be an independent contractor with the New Mexico Museum Foundation. Thomas and Carroll were in constant motion on ventures in Nepal, India, and Tibet.

My activities centered on housekeeping, finances, serving on the bank board, attending to Cow Creek, playing squash, hiking, and entertaining—and, on occasion, being fed by—my several lady friends. I lost several good friends in Santa Fe and New Mexico. Ralph Burch, once active in Gross Kelly & Co. and later a civic leader in Albuquerque, died on June 1. Julie Paltherique Dougherty, of the Hubbell Indian Trading family, died in June. A great loss to me was my inherited Labrador dog, Tika, who had been my constant companion since Jeanne's death. Lastly, New Wood, owner of Santa Fe Motors and a devoted citizen of Santa Fe, died on July 22.

I took several short trips during the year. In March I spent a week with the Downers in Florida, a welcome reunion for two, now aging, college roommates and longtime friends. In June I joined Joyce Peters for the hundred-year ownership celebration at her family's San Simon Ranch east of Carlsbad. On August 31 Nathan Greer and I joined Yei and Frank, who were at the Broadmoor Hotel in Colorado Springs celebrating fifty years of marital bliss and trauma.

On a more ambitious note, I decided to join the Archeological Institute of America's trip to China. I was apprehensive and somewhat reluctant to take this leap into the unknown, but I had spent one year in Korea facing Chinese troops, and I was curious to learn more about this important part of the world. The itinerary included Beijing, Xian, Dazu, Chongqing, a Yangtze River cruise, Hangzhou and Shanghai.

The highlight of Beijing, the nation's capital, was Tiananmen Square and the Forbidden City, the ancient base for Chinese emperors. Xian, the capital of the Tang Dynasty, contains the 2,300-year-old site of the terracotta warriors guarding the grave of the first emperor of China. Complete military units, detailed down to individual weapons, horses, and materiel are on parade, yet only 20 percent of the site has been excavated.

Our four-day cruise down the Yangtze River brought home for me the sheer magnitude of the massive Three Rivers dam project. Construction was just starting. One of the primary reasons for building the dam was to control the wild and widespread flooding of the river. Filling the reservoir behind the dam would require resettlement of over five million people. The electrical power generated by turbines was to furnish 20 percent of the electrical power needed by the nation.

Shanghai was the last attraction we visited on the trip. The colonial waterfront buildings along the Bund reeked with the architectural vestiges of former colonial capitalism. Across the Huang Pu River lay the new and handsome skyscrapers of the "new" China, yet bicycles were still the primary mode of transportation. I found the multitude of people to be suffocating, especially on the night of a holiday in the city. The streets were solid humanity. I was overwhelmed by the thought of supplying these multitudes with food and other necessities and by the market potential of this demand and the productive potential of the population. A disciplined, educated and directed China was obviously a forthcoming reality.

By the end of the year I had had my share of traveling. Only Pamela came for Christmas. She hosted a party for sixty people at my house on December 22. We spent a quiet Christmas together. On December 28, with a blanket of new snow, we cross-country skied from Leland Thompson's La Bajada Ranch to the Santa Fe River gorge. On December 30 I took Pamela to the Albuquerque airport for her return to San Francisco. On New Year's Eve, Mary Redmond and I celebrated the New Year at Andy and Kay Black's party at Nambé.

As usual, my 1997 resolve of good works not performed was paved with good intentions. Maybe in 1998.

1998: Travel and Trauma

My son Robert's art show at the Bentley Gallery in Scottsdale, Arizona, launched the year off to a good start. Since the gallery was relatively close to Santa Fe and Arizona's warmer climate was inviting, I decided to attend the opening. A further impetus was Joyce Peters's invitation to join her in Phoenix, where we were to be house guests of her friend Joan Schneider, a school teacher and former college chum.

Scottsdale was new territory for Robert. The Bentley Gallery was well known to the affluent snowbirds that migrated to Arizona for the winter. He exhibited thirteen pieces priced from $6,500 to $22,000. Robert was pleased with his paintings and enjoyed a warm relationship with the gallery owners. The evening opening on January 7 was well attended. Robert was at his charming best, and Shirine, his Iranian girlfriend, added a sophisticated aura to the gathering. I enjoyed being featured as the father of the talented artist. Nine of his works were purchased and the evening ended with a gala dinner at a elegant restaurant nearby.

When we returned to Joan's house, an electrical power outage plunged the neighborhood into darkness. Joyce and I were not familiar with the house and had planned to share the guest room, which was furnished with twin beds, next to the house's single bathroom. Groping around in the dark with two friendly but not intimate ladies was not my idea of eroticism. We finally completed the bedtime requirements and retired to our assigned locations. The morning dawn revealed us looking somewhat frowzy compared to our appearances at the previous gallery scene. So much for the glamour. Pleased with Robert's success but uncomfortable with the housing situation, I returned to Santa Fe. Joyce continued her visit, happy, I am sure, to be relieved of the male baggage.

I started classes at the Valdez art school. I had taken some lessons during my pre-retirement days, and now, stimulated by the Santa Fe art scene and by seeing my son's work, I started up again. I worked hard but was frustrated by my lack of talent. I could not master perspective. Further, my attempts at capturing the beauty of the nude models did them little justice. I persisted, but enjoyed chainsaw logging at Cow Creek much more.

Another activity I became increasingly involved in was organizing and collating business history, family letters, records, and photographs to give as additions to Gross Kelly & Co. material already in the Business History

Archives at the University of New Mexico. In February I visited the Downers in Florida for a warm week of beach walking and tennis. Topics of discussion there were the United States' and United Kingdom's bombing of Iraq in punishment for sanction violations, and Clinton's affair with Monica Lewinsky.

March weather in New Mexico leaves much to be desired, and this year I was particularly bothered by the monotonous wind, dust, and dry, cold weather. My birthday was approaching and I knew I didn't want to be home, since my children were scattered and unlikely to join me in celebrating my seventy-seventh. I decided to spend my birthday week in New York City and invited Rosie Stewart, who was widowed and made good company. She flew from Denver and I from Albuquerque and we joined forces at the New York Harvard Club.

Rosie, who was an active docent at the Denver Art Museum, broadened my appreciation of the collections at the Frick Gallery, the Metropolitan Museum of Art, and the Museum of Modern Art. Her friend, Fred Norgaard, who was on the editorial staff of The *New York Times*, toured us through the *Times* building, including the newsroom and the executive offices and board sanctum of this famous newspaper. One memory that stands out was the noise and cramped writing facilities for the columnists and reporters. It was hard to imagine being able to compose in that zoo.

We attended plays and enjoyed good dinners. One play I recall was *1776*. It recreated the rancorous arguments and resulting compromise among the members of the Continental Congress. I find it amazing that unity was finally achieved.

I survived the birthday. Rosie was the right choice to see me through the ordeal. We had a good time. She returned to Denver and I to Santa Fe.

I departed again from my Santa Fe routine in May and June. Ed Brosseau was a member of the California Rancheros Visitadores, a club with several hundred members, most of them horsemen, ranchers, and would-be cowboys. I joined them for their annual encampment at the Janeway Camp, some fifty miles inland from Santa Barbara. The diversions included trap shooting, calf branding, horse racing, and rodeos, interspersed with considerable conviviality—drinking, yarn-swapping, barbeques, and campfire visiting.

I was Ed's guest and Jerry Peters was invited by Ralph Tingle. Our flight in Jerry's Lear jet took only two hours. Jerry, always the businessman, was talking on the telephone during the whole flight. At Janeway Camp, I shared a small, two-cot tent with my host, Brosseau. Almost immediately upon arrival, the rains began and fell almost continuously for the next two days. It was a very soggy and muddy time. Most of the planned activities were slithering comedies, rather than demonstrations of feats of horse- and cattle-handling skills. Of course, the main covered meeting area became a fulltime bar scene.

The rain spared the cannon demonstration. Several teams competed to see which could propel a one-foot-diameter bowling ball the farthest. Being an ex-artilleryman, I enjoyed the contest. The jerry-rigged cannons, jammed with black powder and bowling balls, were fired in sequence. The balls hurtled into the air and landed some hundred yards down range. The competition was intense and exciting. John Bokum's team won.

This Visitadores event was the wettest on record. I was glad when it ended.

In early June I attended my fifty-fifth class reunion at Harvard. The class ranks had thinned even further. Memorials to deceased classmates dampened the festivities. However, the excitement of the graduation ceremonies and the awarding of honorary degrees in the Harvard Yard were, as usual, emotional experiences. The highlight of the reunion for me was the off-campus tour of the warship Old Iron Sides, the USS *Constitution* of Revolutionary War fame, which was moored in the Boston naval yard.

This year, all the children summered in Santa Fe for the first time since Jeanne's death five years before. Robert was working with Tiberon in Albuquerque, printmaking for several weeks. Thomas and Carroll, with their son Liam, had the use of a loaned house and were completing a book on Tibetan language and culture, *Tibet: Reflections from the Wheel of Life*. Pamela was in her new job with the New Mexico Museum Foundation and was now based in Santa Fe. Susan and Christy were active in their real estate ventures. Family gatherings and parties with their friends, along with picnics and visits to Cow Creek, were frequent.

In August Susan became very ill and was admitted to St. Vincent Hospital with severe inflammation of the liver. Requiring intensive care, she was transferred to Lovelace Hospital in Albuquerque, where her condition became life-threatening. On August 31, in dire need for a liver transplant, she was transferred to the University of New Mexico Hospital. She qualified for a transplant; with only 5 percent of her liver functioning, she was near death. Miraculously, a donor was located and on September 1, after a six-hour procedure, she was saved by the heroic and skilled liver team of Drs. Bijan Eghtesad and Hecker.

Although we were jubilant over her good fortune, Susan's convalescence did not progress properly. Apparently, there was an arterial blockage of the blood to the new liver, and it failed. On September 21 she received a second transplant. The substitute liver was tolerated and Susan appeared to begin a proper recovery. Then, shortly after the operation, her left hand became discolored. The condition worsened and it was determined her hand had been deprived of oxygen during the liver transplant. She faced the possible removal of fingers from her left hand.

On October 20, doctors operated on her hand, removing the little finger. On November 9 Susan was finally recovering at home and I took her out for dinner November 20. In another ten days, she was permitted to drive her car.

Susan's ordeal was traumatic for her family and certainly for me. I commuted to and from Albuquerque constantly. At her home, her young children and housekeeping required continuous attention. Once again, Christy was a model of devotion and levelheaded competence. As Susan recovered, the real meaning of Christmas became evident. Susan was back with her family and was mending well.

1999: Le Fin de Siècle—Relative Calm

The final year of the twentieth century passed with relative calm. No severe crisis or tragedy occurred. My children, who were all born within this cataclysmic century, were reasonably well poised to face the challenges of the twenty-first century.

Susan was recovering from the nightmare of her traumatic liver transplant and its complications. Though still frail, she was coping with the challenges of raising a young family and maintaining a happy yet volatile marriage.

Pamela was managing to navigate the rapids of a new career with the New Mexico Museum Foundation as director of licensing and product development. At the same time, she was adjusting to the demands of predatory males who were pursuing an attractive, recently liberated woman.

Robert's reputation as a gifted and dedicated artist continued to grow. From his New York City base, he supplied galleries in Santa Fe, Scottsdale, Sun Valley and San Francisco.

Thomas and Carroll, the Nepali expatriates, were seasoned chroniclers of the cultural life in Nepal, Tibet and India.

At the age of seventy-eight, I was really the luckiest of men. I enjoyed good health, was financially secure, and remained engaged in many diverting and absorbing interests. Even with these undeserved gifts, though, I tended to seek out problems to worry about. I was always concerned about the wellbeing of my children. I questioned the sanity of living alone in a large and somewhat isolated house. The reality of death, the disposition of assets, and the avoidance of excessive estate taxes—these preoccupations were my mental companions.

Early in the year, my brother Booker's heart condition recurred, which led to another open heart operation. He was hospitalized in San Francisco, where the lower altitude reduced the demands on his weakened and oft-repaired heart. After a long recuperation, he resumed his now quiet and limited life in the Santa Barbara area of California—quite a change from practicing law for over forty years in the high altitude of Northern New Mexico.

I spent my seventy-ninth birthday with Pamela at the Tanque Verde guest ranch near Tucson. Jeanne and I had taken Pamela there thirty-nine years before when she was twelve. Fortunately, the horses we rode were not the same ones. The desert around the ranch featured interesting trails and it was fun playing cowboy again and enjoying tennis, birding, and the warm weather.

After the March birthday jaunt, Joyce Peters invited me to her nephew's wedding. The agenda was enticing: a tour starting in Virginia and heading southward, ending with the wedding in Mobile, Alabama. The proposed trek would immerse us in the geography and culture of the South, a region dear to my Confederate heart. Despite some concern over the length of the sojourn, I readily accepted the invitation.

Joyce's sister, Mary Stokes, lived in Upperville, Virginia. Her son, Claiborne Stokes, was marrying Clara Leder Lamar, a Mobile, Alabama, belle. Our trip started at Mary Stokes's home in a region renowned for its gentleman farmers, beautiful pastures, spacious homes, horse breeding, and hunting with hounds. Here I met Claiborne, the groom, an attractive young man who was a history buff and knew much about Civil War skirmishes and cavalry engagements in the area.

From Virginia, we drove to Washington, D.C., where we paid a visits to the Korean and Vietnam war memorials. The Korean memorial disappointed me; perhaps its depiction of poncho-draped soldiers picking their way carefully through a muddy field was too realistic and not romanticized enough for me.

After we visited the charming University of Virginia and the Jefferson plantation at Monticello, I understood why Jefferson was always in debt. It seemed to me the overhead costs of supporting more than one hundred slaves and keeping up the splendid estate would have far exceeded any income he could have derived from farming.

Our route south led us to Norfolk, Virginia, then along the Blue Ridge escarpment into Ashville, North Carolina. Traversing the Appalachian range was a revelation for me. Seeing it, I could understand what an awesome barrier it was for settlers migrating westward. We crossed the Great Smoky Mountains into Tennessee through the homeland of the Cherokee people. This tribe was an unfortunate pawn traded back and forth between the British and the Americans and finally forced to march along the Trail of Tears to resettlement in Oklahoma.

My part-time Santa Fe neighbor Jim Baker, who maintained his primary residence in Chattanooga, toured us around that fair city. From there, he drove us to the nearby site of the bloody Civil War battle of Chickamauga. The Union army fought here to capture Chattanooga and take control of the navigation of the Tennessee River, while the Southern army sought to defend their hold on the area for the same reasons. Joyce Peters's great-great-grandfather, Confederate Colonel Claiborne Merchant, played a part in this battle, won by the Confederates. Later in life, Merchant was a major cattle baron in Arizona and New Mexico.

From Chattanooga we entered Alabama and stopped in Moundville, the location of earthen mounds erected, apparently for ceremonial purposes, by the original Indian inhabitants of the region. I had learned of these ambitious creations from my archeologist friend Steve Williams, who had excavated some of them.

When we arrived in Mobile, the post-bellum matrimonial ritual of the prominent Southern bride was in full swing. The wedding party included twelve maids and groomsmen. Mary Stokes hosted the rehearsal dinner at the Mobile Country Club. Everyone was decked out in formal attire. Long and flowing toasts were interspersed between courses during the lavish dinner. The affair lasted well into the night and was followed by a formal ball. For me, the experience was shades of *Gone With The Wind*.

The formal nuptials were held at the Episcopal church, followed by a reception, again at the Mobile Country Club. Fortunately, my duties as an escort were minimal, enabling me to participate as an interested and cordially entertained observer.

During the Civil War, Mobile was a key supply point for the Confederacy. The town's port facilities were excellent. The entrance to the Mobile Bay harbor was fortified with chains to prevent hostile ships from entering. The Union Navy, commanded by Admiral David Farragut, finally breached the Confederate defenses when he issued his famous order, "Damn the torpedoes, full steam ahead!" The last battle of the Civil War took place in Mobile. I found the spirit of the South still evident at the time of my visit and I returned to Santa Fe a romantic Confederate—thankful, however, that the Union was preserved.

Once home, the realities of every day living soon dispelled the romance of the trip through the South. I had to face deferred maintenance projects at the Cow Creek property. The roadway had suffered erosion, and fallen trees and new growth had restricted easy passage. I engaged Joe Archuleta, who owned a vintage road grader, to manicure the neglected roads. The work took a number of days in on-again, off-again fashion, as rain muddied the area often and prevented grading. Also, Joe's grader suffered from old age and needed several mechanical transfusions. The job was finally completed and permitted the annual "erection day" — the raising of the tent on its platform and outfitting it for summer occupancy.

At home on Palace Avenue, my neighbor Jim Baker drilled an irrigation well adjacent to my property. He cordially suggested we jointly install the well and irrigation system for mutual use. I spent considerable time researching the problem. Unfortunately—or perhaps fortunately—my sprinkler installer forcefully discouraged me from joining in the project, warning of disputes that could arise from joint ownership. I decided to forgo the venture and instead continued the archaic and expensive task of using conventional city water for watering my property.

During this "maintenance period," I became interested in the idea of moving to El Castillo, a residential facility geared to the comfort of older persons. At the time, El Castillo was building a new, several-story facility beside the Santa Fe River. I inspected a penthouse unit somewhat isolated from the rest of the units, with a view of the town and part of the nearby mountains. The idea of living in a new building, eating in a dining facility, and, most of all, having no responsibility for upkeep and maintenance, attracted my fancy. After my investigation, I reserved the penthouse and deposited a substantial down payment on the project. A final determination and full payment was required in September.

I struggled almost all summer over the decision about El Castillo. I interrogated friends and residents. I counseled with my children. I reviewed finances. I was really a complete bore because of my indecision, but finally I realized I really did not want to move. I was still vital and physically active, and I had to admit I enjoyed the challenges of independent living. Shortly before the September deadline, I voided the reservation and felt relieved.

Despite the mental gnawing of indecision, I kept active all summer, working at Cow Creek, taking drawing lessons, playing squash and tennis, and continuing some pro bono work. The children came and went, and during their visits the family problems and relationships kept us closely knitted.

During one visit with Pamela, she expressed an interest in visiting some places I had told her about but which she had never seen: Hermit's Peak, Tecolote Peak, Bernal Peak (Starvation Peak); the villages of Anton Chico, Chapelle, Dilia; and the old Gross Kelly cattle ranch. These points of interest were all in the Las Vegas, San Miguel County, area. We climbed Hermit's Peak and visited the hermit's cave, then visited Ojitos Frios next to Tecolote Peak and ascended that peak, which stands beside the Santa Fe Trail. From the flat-topped summit of Tecolote, we took in the vista of the eastern plains and mesas beyond Las Vegas—a considerable spread of interesting country. Starvation Peak demands a rough climb to its rim-rock summit. Once the refuge of local inhabitants fearing raiding Indians, in more recent times it marked the northern border of one of the Gross Kelly ranches. We climbed it, too, and ended our trip of memories by visiting the isolated and quaint lower Pecos river villages of Anton Chico and Dilia.

I spent a week in Washington, D.C., the guest of Victor Delano, a retired Captain of the U.S. Navy. Shortly after World War II, Victor was assigned as a naval liaison officer to the Los Alamos Scientific Laboratory. We became friends and maintained the relationship. I stayed at the Metropolitan Club, near the White House. Victor was now a widower and owned a spacious apartment in the Chevy Chase area. Being an active alumnus of the Naval Academy, he hosted me to an Annapolis-Tulane football game. Navy won.

While in Washington, I visited the White House, paid my respects to Senators Domenici and Bingaman and toured the Supreme Court and the Capitol. Another highlight was a private tour of the vast Pentagon. I marveled how inter-service coordination could be maintained in such a vast complex.

I spent the last few days of the century getting ready for Christmas and for the annual Senior National Squash Tournament at the Kiva Club. The coming of the new century made some of us older players fear our playing days might soon end.

A Christmas present arrived in Nepal on December 26: the birth of Thomas and Carroll's second son, Gaelan.

One sobering aspect of what otherwise was a relatively pleasant year was the death of two of my close friends. Dr. Phillip Schutz, long-time physician, hunter, hiker, and a St. Francis to wounded animals, died. His loss was a severe blow to me. He had been my doctor, counselor, and hiking mentor.

Also, my college friend Stan Durwood of Kansas City failed to make the cut. He telephoned me and said goodbye two weeks before his death. He was a successful movie theater mogul and a strong supporter of youth activities in Kansas City.

December 31, 1999, was the end of a momentous century in which I played a minor yet extremely lucky and active part. I wondered, "What does 2000 have in store?"

Jeanne and Bud Hosting a Garden Party, n.d.

Tented Camp at Cow Creek, n.d.

Ski Break—Pam, Bob, and Tom at Cow Creek, n.d.

Gary Valencia, Fire Cleanup, Cow Creek, June 3, 2000

Bud and Jeanne on a Cruise, 1980

Margaret and Dan Kelly, Bud's Parents, on Their Fiftieth Wedding Anniversary

Caroline Kelly, South Pacific Red Cross, WWII

Kelly Siblings, Booker, Mark, Bud, and Marie Ellis (Yei), n.d.

Booker Kelly and Susan Harrison Kelly, n.d.

Kelly Family, All Together, " The Whole Enchilada," ca. 1991

2000-2011

SEIZING THE DAYS

2000: Cuba and France Revisited

Despite the forecasts of doom and the prospect of the world's computers going crazy, the new century commenced without major incident. There was little change in my routine. I continued with my weekly hikes, lunch at the Kiva Club on Mondays, and monthly meetings of the board of the bank. I played squash regularly and took part in evening dinner/discussion sessions with the Chili Club on the third Monday of the month. Depending on accessibility, I ventured to the Cow Creek property and, if the snow pack permitted, I burned slash and brush we had piled during the previous summer.

I continued organizing Gross Kelly & Co. records and reports I felt were of historical value. I also gave some attention to my sister Caroline's letters and scrapbooks covering her activities before and during World War II. This effort included reviewing my father and mother's numerous letters and recorded activities, as well as their estate papers, which painted a broad picture of the century past. I planned to add the collections to the family archives already on file at the University of New Mexico.

With the advent of my seventy-ninth birthday in March, I decided to join the Academic Research Seminar tour of Cuba, organized by the Council for International Relations (CIR) and led by Cuban-born Professor Nelson P. Valdez of the University of New Mexico. Our group of twenty-five flew to Cancún, Mexico. After a chaotic layover at the airport, we boarded a Mexican airline plane and flew to Havana for a twelve-day tour concentrated on researching Cuban education and health facilities. Traveling by airplane and bus, our group visited Havana, Santiago, El Cobre, Camagüey, Trinidad, and Varadero.

Without exception, the Cuban officials, guides and the general population were friendly and anxious to impress the group with the progress made since the revolution. In my opinion, the two outstanding areas of advancement were in primary education and basic health. Literacy had advanced from practically zero to well over 80 percent, and the same percentage attained a minimum of an eighth-grade education. Cuba was training and exporting a surplus of well-trained primary practice physicians. Furthermore, in urban areas, block teams of doctors and nurses provided basic healthcare to the population.

In economic terms, though, Cuba was hurting. Russia had helped subsidize the economy from 1972 to 1985, providing low-cost oil in return for Cuban sugar. While this trade lasted, conditions were favorable. With the breakup of the Soviet Union, however, Russia abandoned Cuba, leaving it essentially bankrupt. Economic conditions during our visit were dismal. A diet of rice and beans was the usual fare, except in tourist facilities. Parts for machine repair were lacking and the use of bailing wire was the norm. But despite these conditions, the population appeared resigned and generally cheerful, helped by widespread neighborhood music and dancing. European and Canadian tourists luxuriated on the gorgeous beaches of Varadero and stayed in sumptuous hotels financed and operated by European and other non-U.S. corporations.

I think most of us on the tour admired the Cubans, if not their rulers. We were impressed with their friendliness and, at the same time, with their dignity and acceptance. They appeared proud of their relative independence from American opulence and glitz. Our group was in favor of removing the sanctions imposed on Cuba by the United States.

In contrast to the relatively serious academic visit to Cuba, later in the year my twin sons joined me on a sentimental journey to the part of France where I served during World War II. Thomas flew from Nepal and Robert and I flew from New York City for what was to be a deluxe trip, intent on enjoying a reunion of three related men in la Belle France. Robert, the renaissance man, had made all the reservations for this fifteen-day adventure.

Thomas arrived in Paris before us and was staying with the French midwife who had delivered his sons in Kathmandu. When Robert and I arrived, we proceeded confidently to the elegant West End Hotel, but upon our arrival, Robert was advised that we had no reservations. He had failed some way in making the necessary transatlantic confirmation. Paris was jammed, with the European Champion Soccer matches and the French Open tennis tournament in progress. Despite the best efforts of the hotel management, no alternate space was available anywhere in town. We both were exhausted from the flight and in foul moods because of the reservation mix-up. In desperation, Robert called Thomas, who was awaiting our arrival at his midwife's flat. Robert relayed our predicament. Thomas conferred with the midwife. He called back in a few minutes with the welcome news that we now had a reservation at the Garden Hotel in the Thirteenth Arrondissement. The Garden Hotel had no star rating and the price per night was only forty francs. Rich beggars can't be choosers.

Thomas met us outside of the hotel. The proprietor, Monsieur Adam, greeted us in his undershirt. We hauled our considerable baggage up several flights of narrow wooden stairs to our rooms. There were no flowers, no view—just narrow beds, bare closets, and modest bathrooms. And so our luxurious tour began.

As it turned out, our accommodations were adequate. Monsieur and Madame Adam became our friends. The savings were fantastic and when we returned to Paris, we enjoyed our stay at the Garden Hotel and the friendly simplicity of our innkeepers.

The region we planned to visit was generally in the part of Normandy and the Atlantic coast of Brittany the Allied forces liberated. Our first excursions, however, were in Paris. A river trip on the Bateau Mousse oriented us to central Paris. Thomas had a friend, Maria Caldera, who worked at Cartier, the famous luxury jewelry establishment. She suggested restaurants and points of interest we should see. She also invited us to her apartment for an initiation dinner.

Our second nighttime excursion in Paris lasted all night. It began with our dinner at Maria's, graced by the presence of one of her lady friends. The dinner, with lots of wine, stretched into a dance marathon. The two women were nimble dancers and we three, well-lubricated males thought we were Fred Astaires. At dawn, our French charmers finally were liberated from us and we maneuvered ourselves back to the Garden Hotel to sleep and then resume our tour of Paris.

Leaving Paris, we headed for the Normandy coast. Our route led us into the area of Caen and Bayeux the British invaded during the war. Just seeing the famous Bayeux tapestry, which depicts the Norman invasion and conquest of Britain in 1066, made this trip worthwhile. From Bayeux we drove to Omaha Beach, where American forces had struggled on to French soil during the D-Day invasion. There was still enough battle evidence left to swamp our minds with images of the peril of that invasion. Thomas and Robert were visibly shaken by imagining what took place some fifteen years before they were born.

We climbed the rugged cliff at the Pointe du Hoc and entered the German fortifications that finally, after much bloodshed, were taken by the U.S. Rangers. Visiting the U.S. cemetery, directly inland from Omaha Beach, was the emotional climax of our day. We were deeply impressed with the sight of ten thousand headstones, each with its own miniature American and French flags bravely waving beside it. The green grass, the white headstones, the cloudless day, and friendly greetings from the French locals—it all stamped us with a lifetime memory.

From Normandy we drove to Mont St. Michel, the famous abbey on the boundary between Normandy and Brittany. The tide was out, so we were able to make the somewhat strenuous hike to the top of this striking landmark. We then drove across the Brittany peninsula to the port city of Lorient on the Atlantic coast.

My main objective in coming to France was to visit Lorient. In 1945 I had viewed it for several months through the window of an L-4 artillery observation airplane. At that time, Lorient was a key departure and refitting base for the German submarine "wolf packs" that preyed on the Allied ship convoys steaming from America to England. A pocket of some forty-thousand German troops defended the port and surrounding area, which was bypassed by the Allied invasion forces heading for Germany.

Our division, along with an odd assortment of French troops, had the mission of containing this pocket of German forces, interdicting German submarine traffic and defusing the base. In 1945 Lorient, with the exception of the submarine bunkers, was little more than a pile of rubble under constant artillery and aerial bombing attacks. Now, in 2000, it was a bustling city of some two-hundred-thousand inhabitants. The only remnants of the wartime siege were formidable cement bunkers that had been refitting facilities for the submarines. They were a constant artillery target in my wartime day. It was an eerie feeling entering these cavernous workplaces, now used by French ship-repair shops.

From Brittany we returned to Paris and the charming Garden Hotel to resume our explorations of the city. We visited the museums, Notre Dame, Les Invalides, and Napoleon's tomb, and sampled several excellent restaurants. We repaid Maria by taking her to the restaurant of her choice. On June 6 Robert took me to Charles de Gaulle airport and saw me aboard the airplane to New York. Both boys had continuing business in Europe and I had heard vaguely distressing news of events in New Mexico. The revisiting was over and it was time to head home.

Pamela met me in New York with the traumatic news of the Vivash fire and the total devastation it had wrought on our cherished Cow Creek property. The fire, which was apparently human-caused, broke out on May 29, Memorial Day. The conditions were perfect for a conflagration. The area was parched, the winds high and the country overgrown with excessive trees and brush. The wind drove the initial blaze from high on Vivash Mesa into Cow Creek canyon. From there it roared northeastward, following the contour of the canyon. It engulfed our property and then headed ferociously up the canyon, eventually consuming everything in an area of some thirty-thousand acres. Ultimately, the Forest Service controlled the rampaging beast and rehabilitation slowly began.

Access to the property was hindered by flooding following the rains later that summer. The moisture was welcome, but the denuding of the canyon slopes was so complete that Cow Creek was bloated with fire debris and it overflowed, rendering the somewhat primitive road impassable until crews cleared it.

Once the route was passable, Wooly Henry, my faithful "Sherpa" Gary Valencia, and I were able to reach the property and survey the damage. It was a wasteland. With the exception of two small, widely separated clumps of conifers, everything was dead. Standing trees had been suffocated, all underbrush was gone, and the whole area was one blackened, ash-covered scene of desolation. It looked like a battlefield. All that remained of the fences was naked, scorched posts; the water hydrants were heat frozen. The only relic of the ambitious tent platform was the line of nails that had once held the wooden deck in place. The burned water line and electrical lines were exposed by erosion. The paths and roads were covered with a piles of burned trees and forest debris. It was total destruction.

I felt a precious part of my life had been violated. Of particular poignancy was Jeanne's grave. I had buried her ashes, as well as those of her beloved dog Tika, in her favorite, private grotto. It was a small oval-shaped patch of ground backed by lichen-covered boulders and surrounded by beautiful aspen trees. The savage fire had consumed the lichens, leaving black, naked boulders that seemed to leer out at the treeless, ashen landscape. The only vestige of the sanctuary was the small silver crucifix cemented to the rock outcrop.

The emotional impact of this fire was severe. I was distraught, nervous, exhausted, and depressed. It also impacted my family. Nevertheless, I began the process of rehabilitating the property: reseeding, harvesting the fire-damaged but marketable timber, and rebuilding the fences, roads, and utilities. This turned out to be a long and frustrating, bureaucratic ordeal.

Once I cleared the hurdles of environmental, archeological and endangered species regulations, I was able to negotiate a timber contract with Rio Grande Forest Products, Inc. Bill Heath and his crew from Gallina, New Mexico ably cut and delivered the timber to the company's mill in Española—in all over four-hundred truck loads of merchantable timber. It was awesome to see these huge trucks, loaded and making their way on the steep and narrow logging road Bill constructed on the property. When the logging operation was completed, the salvaged timber sales totaled about thirty thousand dollars.

While the logging was in progress, I hired the Valencia family who, with the help of friendly volunteers, cleared and stacked brush. I reseeded most of the property and, when there was ample snow pack, burned the piled slash. This prepared the property for nature's rehabilitation.

In November Susan confided in me that she was overwrought with domestic strains. Wisely, she enrolled in a program offered by the Sierra Tucson center. The climax of the program was a Family Confrontation Seminar, during which Susan voiced her problems involving her family, her siblings, and me. In turn, family members aired their feelings. It was a wrenching but extremely constructive process that helped establish behavioral boundaries, expose destructive habits and devise remedies for underlying problems that had been smoldering for some time. Not unlike the Vivash Fire, the buildup in family stresses had erupted in an explosion. The family outpouring initiated our rehabilitation.

Informed and strengthened, we returned to Santa Fe a week before Christmas. We all felt purified and ready for the healing that would continue in the New Year.

2001: Penance and Absolution

The emotional impact of our family week at Sierra Tucson influenced my life throughout the year 2001. I acknowledged the character revelations that surfaced during the confrontation. Among them was my impatience and temper when imbibing at family gatherings. I also realized I tended to be judgmental, to lack self control, and to communicate my thoughts and feelings poorly.

In an effort to overcome these shortcomings, I enrolled in the Al-Anon program and attended meetings where participants discussed their experiences and problems. Simply airing concerns reduced trauma. Members showed their sympathy and quite often suggested remedial action.

Undoubtedly, I had been in denial. I admitted my shortcomings but did not feel I was an "enabler." I enjoyed social drinking but did not think I had a serious alcoholic addiction. In any case, I went "on the wagon" and, with some exceptions, generally abstained throughout the year.

In an attempt to bring family members into a closer and more harmonious relationship, I invited them all to be my guests for a visit to Cayman Island, a British colony isolated in the middle of the Caribbean. Tom and Carroll and their two boys traveled from Nepal, halfway around the world, while Robert and Shirine flew in from New York City. Pamela and I, as well as Susan and Christy and their two children, the balance of the tribe, departed from Albuquerque. We rendezvoused in Miami and then flew over Cuba to Cayman.

In addition to beach walking and swimming, golfing and relaxing, we took excursions, one of them on a chartered boat to a shallow offshore location where stingrays congregated. Although they are capable of delivering severe stings, here the rays were docile. We walked through the shallow water and could mingle with and even handle the ugly creatures safely, as long as we kept both feet on the ocean floor. No one was stung. I was relieved, however, once all were back safely in the boat.

My birthday was part of the reason for the Cayman gathering. Another was to comply with a recommendation suggested at the Sierra Tucson mediation and reiterated at Al-Anon meetings: to seek reconciliation and forgiveness for past grievances. In my case, I felt the need to seek expiation from my son Thomas. As a child, he annoyed me and I typically reacted with impatience and anger. During his adolescent days, I sometimes found him unattractive and boorish. I was critical of his outlook and behavior and frequently berated him unnecessarily. His early departure from home and long exile in Nepal was, in my mind, partially my fault. I felt the need for a general confession.

I had prepared for the occasion. I invited him to join me on a long walk on the isolated Cayman beach. The confession came as a surprise to Thomas. He had not felt a deep need for my apology. However, my candor warmed him. Any underlying resentments we may have harbored were wiped out by this session. On returning from the hike, we both felt our relationship was reborn. Cayman Islands are headquarters for offshore banking; in my case, I had made a profitable offshore deposit.

For most of the summer and fall, I was an almost daily visitor to the Cow Creek property. Friends and I began hand broadcasting grass seed on the previously logged areas. We packed fifty pound sacks of assorted fescue, grama, wheat, and rye seeds into backpacks and trudged with them to designated areas, where we spread the seeds. This usually was a weekend venture.

Concurrently, the destroyed boundary fences were rebuilt. This involved clearing the fence lines of fallen trees, resetting the metal fence posts, and stringing new wire, usually on steep grades. Over a mile of four-strand fence line was laid. The buried water and electric lines that had been exposed by erosion also needed replacement. This entailed digging four-foot-deep trenches and laying and burying new pipe and electric lines.

By the time most of these chores had been accomplished, cold weather was setting in. Urban wood-burning stoves and fireplaces needed firewood, and I had it in abundance. Most of the standing timber consisted of small, noncommercial dead trees, fallen logs and broken and scattered branches that also made ideal firewood. I needed to clear these out so I could reseed and replant, so I sought to develop a market for this mass of potential firewood. This meant I had to find harvesters, wood dealers, and consumers, and time was a critical factor. Once snow came, the roads would be closed and nothing could be harvested or delivered. Thus, I needed paid and/or volunteer firewood harvesters, but workers were hard to find, especially volunteers. The harvest area was remote, making access a challenge for many people. Once the trash timber was gathered, it had to be cut to lengths suitable for fireplaces and wood stoves. Lastly, the wood had to be sold and delivered to the user.

After a series of inquiries, I found two interested wood buyers—La Fonda Hotel, via my friendship with the hotel owner Sam Ballen, and Hotel Santa Fe, via its manager and my fellow squash player Paul Margetson. Turning to my dependable "Sherpa" family, I made a deal with Gary and Eloy Valencia. The arrangement tilted strongly in their favor. I supplied the wood source and the buyers. They delivered the wood in Santa Fe and pocketed the payment. It was a win-win situation for all. The arrangement with La Fonda lasted through the winter. However, the hotel housekeeper objected to the blackened bark on some of the wood, which was messy and complicated the task of keeping the hotel rooms tidy. Sam Ballen understandably accommodated his housekeeper. Paul Margetson used the wood in the main lobby fireplace, so the blackened bark was no problem; he remained a customer until the Valencias became tired of making the long haul.

By the end of the year, Cow Creek was healing well. Nature was re-clothing the land with aspen sprouts and grass. The slash cleanup became an in-house stack-and-burn operation, which still continues.

Squeezed in between these and other activities, I enjoyed a weeklong return to the Civil War area when Joyce Peter's niece, Stormy Stokes, was married in Upperville, Virginia, on May 19. I was again asked to be Joyce's unofficial escort, as was the case for the Australian wedding.

We were houseguests of the charming Virginia lady, Ann MacLeod, whose spacious home was known as Dunvegan Farm. She had an intimate knowledge of the Civil War battles fought in the area. Informed by her insight, we visited Harpers Ferry, where John Brown was apprehended prior to the war. It was important as the federal arsenal, where the army's rifles were manufactured. The Confederates, commanded by General A. P. Hill, captured the arsenal.

Our next visit was to Antietam National Battlefield, site of one of the bloodiest battles of the Civil War. Immediately after taking Harpers Ferry, General Hill marched his troops seventeen miles to Antietam to rescue General Robert E. Lee's Virginia army from near defeat by Union forces under General McClelland.

We returned to Upperville and the marriage ceremony, which took place at the graceful Episcopal church, built by Paul Mellon. The day following the wedding, we continued the Civil War tour, visiting Gettysburg, where General Lee's march on Washington, D.C. was blunted, resulting in over forty thousand casualties. Lee retreated across the Potomac, and the battle site became history.

The wedding and the battle tour were an unforgettable experience that abruptly ended on the return to Santa Fe on May 21.

A tragedy struck on June 3. On that day, the crown prince of Nepal murdered his parents, the king and queen, and then killed himself. Thomas and Carroll were close friends of the prince and of his sweetheart, whom he had not been permitted to marry. They also knew the princess who was slated to be his bride. Besotted and frustrated, the crown prince ended a dynasty. The king's brother, whom Thomas did not respect, became the heir apparent and the new king. Tension was high and Thomas was concerned that uprisings might occur. Obviously, his anxiety was relayed to us. The crisis subsided and we in far-off Santa Fe were relieved. Thomas and his family resumed their abnormally normal lives.

The month of July was marred by the death of Frank Gorham, my brother-in-law. Also, during the month, I served as proxy father of the bride at the wedding of Seonaidh Davenport and Brian Sharon, because the bride's father, David, was too ill to give her away. David died in September and was buried at the National Cemetery. He was one of my oldest and dearest of friends.

Later that summer, I became interested in another trip sponsored by the Council on International Relations, this time to Japan. Having been there in 1952 during the Korean War, I was curious to see the changes that had dramatically altered the nation during this fifty-year span. I signed on for the tour, which was to run from October 26 through November 8.

On October 29 our plane arrived at Haneda Airport. I had landed there on my way to Korea in 1952. We were bused sixty kilometers to our hotel in Tokyo. The city was drastically changed. It was a twenty-first century star, completely modern. The population dressed in chic Western clothing. The heart of the city, the Ginza, boasted offices of the world's foremost corporations, including the great Japanese business entities that had made this once destitute and devastated nation a world economic leader.

The three-day experience of Tokyo included a stop at the American Embassy and visits to the Imperial Gardens, museums, the convention center, and other points of interest. On our final morning we visited the fish market, the largest in the world, where fresh catches of all varieties were prepared and shipped. Departing Tokyo for Takayama, we rode one of Japan's spectacular trains, which are fast, clean and efficient. Here, we were initiated into the traditional domestic ritual of bathing, dressing, dining, and sleeping. My roommate was eighty-three-year-old John Potter. I was eighty.

Leaving our street shoes at the appointed location in the Ryokan-Hiranoya Hotel, we were escorted to our room by two young women dressed as geishas. They served us tea and then departed. The next assignment was disrobing and being clad in kimonos. Then we proceeded to the male bathing facility. The routine here included extensive soap-lathering, a conventional shower, and then a good soak in a large, heated community tub. Following that, we had another soap-down and shower and returned to our room.

Dressed for dinner in formal kimonos and shod in split-toe slippers, feeling odd and embarrassed, we proceeded to our special dining room. We sat on the floor at small dining tables in two facing rows. The ladies all loved their costumes and were much more at ease than the men. We enjoyed a multi-course meal, liberally lubricated by sake.

Following dinner, fully satiated, we were relieved to rise and return to our rooms, where John and I found two adjoining futons—on the floor. John was probably bothered by my snoring, as I was by his frequent trips to the adjoining bathroom. And so the night passed.

The theme of our tour was art/culture, urban planning, religion and sightseeing. We stayed on Honshu Island, the largest and economically most important of the some three thousand islands making up the territory of Japan. New Mexico has a population of 1.5 million and is two-and-a-half times larger than Honshu. Of the 127 million Japanese, over 35 million live in Tokyo alone, the world's largest city. But despite the density of the population, regional and geographic differences within Honshu were diverse. Our visit took us to quaint, isolated mountain villages, regional religious centers, agricultural areas, and manufacturing complexes.

Somewhat jaded by the intensity and diversity of the tour, we spent our last day in Kyoto, a city of 1.4 million people, one hundred temples, and thirty-six universities. That afternoon we attended a lecture, "How Kyoto will develop as an international city co-existing with world heritage," presented by Kyoto's city planning bureau. My diary notes commented, "The lecture was pretty dull. It was amusing to see our group half asleep."

Once back in rural New Mexico, it was a relief returning to the familiar and the basics. Thanksgiving dinner, St. John's meetings, wood harvesting, squash, weekly hikes, and preparing income tax returns left little time for boredom. Pamela and I had Christmas dinner at the Barker's.

I joined Jim and Ruth Ritchie and their other guests on New Year's Eve. We shortcut the evening by celebrating the new year at 10 o'clock and watching the action in New York via television.

2002: Flight for Life, and Historical Travel

The new year began with a trip to New York and Florida. I had been invited to the annual meeting of the Harvard Club of New York City. As a member for over fifty years, I received a special invitation. Harvard's new president, Lawrence H. Summers (who had served as secretary of the Treasury under President Clinton), was the guest of honor. He was a controversial president and I was anxious to meet him. My main reason to go to New York was because Robert was negotiating to buy a large apartment and needed some financial support. I also wanted to see the site of the 9/11 bombings, by now a pit of destruction being cleared.

The Harvard dinner was a gala affair. President Summers took the opportunity to defend some of his recent, controversial actions and comments at the university. He had taken a lot of flak for disciplining a professor in the Afro-American Studies Department and for statements Summers made (misquoted in the media, according to him) about the limitations of female instructors in the biological sciences. He spoke convincingly and, by the end of his talk, most of us agreed with his actions. (He shook too many established academic bushes and later resigned as president.)

The gaping 9/11 hole was near the apartment Robert wanted to buy. At the time, volunteer firemen from various fire departments throughout the nation were conducting tours of the site. Dressed in their fire gear, they added a sense of connection and realism to our tour of the devastated area.

Robert's proposed acquisition was on Fifth Street near New York University. The five-story building had once been an auditorium and was located near John Jacob Astor's former mansion. Robert was negotiating to purchase the ground floor space, which included a basement. The total area was six-thousand square feet. Its potential as a studio/gallery/living space was mouth-watering. However, the financial requirements were awesome. The purchase price was substantial and the building would require extensive renovations to modernize its interior. I was surprised by Robert's ambition and determination. My innate aversion to risk-taking arose instinctively, even though I did want to help him achieve this seemingly impossible goal. (He ultimately succeeded in acquiring the financing.)

From New York City, I again traveled to Florida to visit the Downers at their winter home on Jupiter Island. The highlight of the short visit was a reunion with our former classmate, Dick Harte. We took a daylong hike and picnicked on the isolated beach on Jupiter Island. A scion of an old Boston family, Dick had just returned from a business inspection of the Ames Plantation, a 15,000-acre environmental heirloom located in Tennessee. I was inspired by his enthusiasm for the projects in play there and by the modest "prudent-man" heritage, typical of Bostonians, that he represented. His work at landscape restoration struck a chord with me because of my comparatively modest efforts to work with nature at my Cow Creek property.

Back in Santa Fe, I was busy and generally enjoying life when, on March 5, Susan collapsed with a dying liver. The attack was severe and life threatening. The Albuquerque liver team that had saved her before had been disbanded. To live, she needed another transplant and a facility professionally equipped to perform the operation. We were desperate. Rejection from hospitals in Los Angeles and San Francisco left us at a dead end. Miraculously, the liver transplant team at Seventh Day Adventist Hospital at Loma Linda, California, agreed to admit her, subject to approval and the availability of an organ.

Susan was air-evacuated on March 7, with Christy and me sandwiched in the jet beside her. She lay on a stretcher while two medics continually monitored her vital signs. We flew for one and a half hours at forty-three thousand feet. The hospital ambulance met the plane and transported Susan immediately to the hospital. The next three days were a nightmare. Susan's husband, Christy was thoroughly interrogated before Susan was approved as a transplant candidate. Miraculously, a scheduled transplant patient who was in less serious condition than Susan graciously deferred. On March 11 the operation took place and Susan's life was saved.

We rented an apartment near the hospital. As Susan's recovery progressed, Christy and I were relieved by alternate family members who stayed at the apartment to be near her. On March 22, like a birthday present to me, Susan was released from the hospital as an outpatient and began living in the rented apartment, assisted by rotating members of the family. On April 14 Christy drove Susan to Santa Fe and was finally permitted to return to her own home. By June she had recovered sufficiently to face the demands required of a wife and mother, which had been handled in her absence by Christy, the ever capable and devoted husband and father.

It had been a trying time for all of us. Susan's recovery was a wonder. Although she was on a rigid regimen of rehabilitation, the rest of us could now essentially resume our normal lives.

About this time, I again picked up my interest in family genealogy. This triggered a trip to St. Louis, my mother's birthplace, for the annual Gross family reunion. Yei and I flew to hot, humid St. Louis and took in the event, which took place at a private park and reminded me of a county fair. We met a mob of first and second cousins and had an interesting and entertaining day.

The rules of the gathering limited the attendees to the direct descendants of Jacob and Caroline Gross, the parents of twelve children, including my mother. Of these twelve, each of the nine who parented children was assigned a booth with a placard listing the names of immediate offspring. Yei and I, the New Mexico contingent, were among the gathering of cousins that numbered close to two hundred.

Following the reunion, Buddy Smith, one of the first cousins, gave us a tour of St. Louis. This included St. Louis Cathedral, where our great-grandfather, Dr. Moses Linton, was portrayed in a mosaic in the cathedral's dome, recognizing his medical aid to the poor of the city. We also visited the cemetery plots where family members, along with their servants, were buried. The three days in St. Louis were informative. I found I had quite a group of interesting relatives, both alive and deceased.

In August Wooly Henry and I decided to take a second trip inspired by Stephen Ambrose's book *Undaunted Courage*, a chronicle of the Lewis and Clark expedition. We had previously traveled to the western terminus of their journey. This time, we chose to trace the expedition's route across the continent in reverse, starting in Washington and heading southeast, downstream, following the course of the Missouri river.

Our first leg took us through Wyoming and into Grand Teton National Park, then to Yellowstone and Bozeman, Montana, where my son Robert joined us. He had just finished an art opening in Sun Valley, Idaho. In this case, three was not a crowd; Robert kept Wooly and me from arguing and served ably as an excellent driver.

Our Santa Fe friend, David Grusin, was in residence at his Raw Deal Ranch near Big Timber, Wyoming. He had invited us to fish Boulder Creek, which ran through his ranch. En route to the Raw Deal, we stopped at the Road kill Restaurant for a lunch, which was about as savory as the restaurant's name. Despite Grusin's cordiality, fishing was poor. The boulders in the creek were slippery and the fish wily.

We continued on to Glacier National Park. Here we saw several grizzly bears and gasped at the spectacular but intimidating country. From Glacier, we headed southeast to Great Falls, where we reenacted a portage of the river and followed the rugged course of the Missouri that Lewis and Clark had to negotiate in 1805. Seeing this formidable challenge validated for us that the expedition was definitely a demonstration of "undaunted courage."

Our next stop was the site of Custer's Last Stand on the rolling hills bordering the Little Big Horn River. We were fortunate to have Dennis Fox as our guide. He started our tour seventeen miles from the battle site. We followed Custer's route, noting the location where he split his forces, a maneuver that was a death knell for Custer and the detachment that followed his command. The Last Stand, on a bare hill in Wyoming Territory, was a major defeat and an ignominious finale for colorful Custer. It was a major victory for the Sioux.

We spent our last night luxuriating at the Broadmoor Hotel in Colorado Springs and the next day returned to Santa Fe.

Having traced Mother's family, the Grosses, while at the cousins' reunion in St. Louis, I now wanted to do the same for my father's parents, the Kelly and Thomas families. In conjunction with this search, I decided also to retrace the route of the Santa Fe Trail to New Mexico. In October I flew to Kansas City, where I was the guest of cousin Ann Steele (nee Kelly). Her grandfather was the brother of my grandfather, Harry W. Kelly; their father, W.D. Kelly, had emigrated from Philadelphia to Leavenworth, Kansas, in the early 1840s. He was a successful businessman in the Leavenworth area and had connections with business interests pushing westward as the railroad replaced the Santa Fe Trail. These connections were instrumental in Harry Kelly's move west and the subsequent Kelly legacy in New Mexico. My grandmother Ellis Thomas was born in Richmond, Virginia, and raised in Leavenworth. Her father, a physician and a Confederate, moved to Kansas after the Civil War. Subsequently, she married Harry and spent the rest of her life in New Mexico.

Ann Steele graciously explored Leavenworth with me. She showed me the family homes and the Kelly and Thomas graves. The Presbyterian Kellys were buried in the Mount Muncie cemetery. Being Catholic, Thomas family members were buried in Mt. Calvary cemetery.

Leavenworth was the key town in eastern Kansas in the early days and was the launching point to the West. Fort Leavenworth was the main military post, as well as the headquarters for the command and general staff school. Its luster, however, was tarnished when Kansas City became the business and transportation hub for the region. To me, Leavenworth was comparable to Las Vegas, New Mexico. Both had seen better days.

I now knew who was buried where and hoped to resume the next family search further eastward on some future day.

As planned, Joyce Peters joined me in Kansas City for our journey along the route of the Santa Fe Trail. In a rented car, we traveled on Route 56, which followed the course of the Trail. Our first stop was Council Bluffs, a quiet, aging agricultural community. Typical of former glory days was the local bank named the Farmers and Drovers Bank. The owners, the White family, had operated it for over one-hundred years.

Next we went through Great Bend on the Arkansas River, then on to Pawnee Rock, a promontory that in trail days served as a lookout over the flat, monotonous, buffalo plains of Kansas. We detoured to Fort Larned, a preserved army post formerly active as a security garrison on the trail. The country by now was treeless and more rangeland than farmland. We went through famous Dodge City, where Joyce's grandfather, Colonel Merchant, had trailed cattle to the railroad shipping point. From Dodge, we headed southwest through a neck of Oklahoma and ended the day in familiar territory in Clayton, New Mexico.

On our last day, we visited Capulin Crater, an extinct volcanic cone that rises some six-hundred feet from the plains. The view from its top includes parts of New Mexico, Colorado and Oklahoma. Our route led on along the arid Cimarron cutoff, dodging the longer route through Raton Pass, and on to Wagon Mound and then La Junta, where we stopped at the junction of the Sapello and Mora Rivers. During Santa Fe Trail days, this was a watering and wagon repair facility on the last leg of the journey. To the west, Fort Union, now in ruins, served as the supply depot for the military garrisons spotted strategically along the Rio Grande.

From La Junta and Watrous we drove to Romeroville and to San Miguel del Vado, formerly the customs and regulatory stop for wagon trains entering Mexican territory. We crossed the Pecos River and proceeded along the trail's route into Santa Fe. Our journey took three days, not several weeks as it had during the heyday of the trail. It was an inspiring trip and one that brought into focus much of the history and lore of this famous travel route.

Normalcy reigned for the balance of the year. I continued along my customary path until Christmas. My children elected not to come home for the holidays, but Robert invited me to share Christmas with him in New York. I accepted and flew to New York City, staying as usual at the Harvard Club. Christmas Day was gray and sodden with rain. We elected to go to the Christmas service at St. John the Divine. It was inspirational, devotional, and echoed with brilliant music. Comparing this glorious celebration to the mass-produced, hourly Catholic services at St. Patrick's Cathedral, I was tempted to become an Episcopalian.

2003: Frustrations and Indecision

Once again in Santa Fe, I greeted the New Year 2003 at Jim and Ruth Ritchie's New Year's Eve party. It was festive. I had little reason to suspect that 2003 would be a year of health concerns and frustrating indecision. I also worried about the financial and physical health of my children and the Iraqi war, which added an international ingredient to my cauldron of anxieties.

At the age of eighty-two, I was spoiled by a history of good health. Whenever I experienced something out of the ordinary with my health, I became alarmed and sought a reason for it and relief from my doctors. If I had a pain or was stiff or disinclined to hike, play squash, or work in the woods, I took this as a symptom of something serious and evidence of the start of my inevitable disintegration. Being so overly concerned may ultimately have been prudent, as it did encourage me to take my health seriously. I dutifully, if somewhat reluctantly, followed the guidelines suggested by healthcare providers.

One of my health problems may have been caused by my attempts to fight a blight attacking the piñon trees on our Tano Road property. The overgrown stands, stressed by drought, could not resist the attacks of pine beetles, which created areas of dead and diseased trees. The supposed remedy was to harvest the infected and dead trees, saving those that were still healthy.

I assembled a force to counter-attack the beetles. The Valencias of Pecos, who were skilled chainsaw operators, felled the dead trees and cut them into manageable pieces. We fed the dismembered victims into a monstrous chipper, reducing them into piles of wood chips. It was exhausting work. I was one of the chipper feeders. We harvested over two hundred trees and cleared the property of this initial crop of diseased piñons.

I must have caught the piñon virus, for shortly after the campaign, my back and shoulders developed severe pain. For several frustrating weeks, I could do little but seek relief from chiropractors, massage therapists, acupuncturists, and conventional doctors. Relief and a complete cure finally came from a cortisone treatment recommended by Dr. Feldman.

During this period of physical inactivity, I became concerned over the problem of aging and dependency. This was triggered in part by the decision made by my devoted friends and neighbors, Laughlin and René Barker, to simplify their lives by selling their home and moving to El Castillo retirement complex. I was again tempted by the thought of living in a place where I didn't have to worry about maintenance, garden care, or security, and where I had access to excellent healthcare facilities. The prudence of their move inspired me. I lived alone in a large home and did my own cooking. Home maintenance and repairs were my responsibility. For safety and security, I depended on city police and, in the event of disability or illness, I would be dependent on my children, who were mostly far away. The Barkers's neat and tidy efficiency apartment had some appeal. The companionship of the El Castillo membership was another attraction, and I generally liked the meals I sampled there as an occasional guest.

When I heard of an available vacancy, I again decided to hedge my bet and purchase a unit, but I still went through a period of tortuous indecision. I needed to decide about making modifications to the apartment and what to do with my house. At times, the thought of communal living with a group of retirees became noxious. The idea of purchasing but not owning an expensive unit, which was basically the arrangement with El Castillo, did not seem like a good investment.

Despite these doubts, I persisted. When I sought my children's opinion, they were patient and sympathetic. Generally, they felt I would not be happy in such a controlled environment. Nevertheless, I went ahead. Remodeling plans were drawn and I engaged a contractor. I made a required down payment on the apartment. Caught up in the momentum, I pressed on as if I was going to go through with the deal, even though I was still torn. To do or not to do. Emotionally, I was not ready to give up any independence.

Finally, when it was time to put the required $143,000 on the table, I made a decision. Encouraged by my son-in-law Christy's intuitive opposition to the move, I withdrew my application. The nonconformist in me won out. I left some money on the table but felt greatly relieved from this long period of fruitless indecision. (Several years later and older, I am still living alone and enjoy my injudicious independence.)

This year I was often involved in the business affairs of my two expatriated sons, one in Nepal and the other in New York City, each claiming Santa Fe as his legal place of residence. Thomas's life in Nepal was somewhat primitive. He used New Mexico as his U.S. financial headquarters and me as his business agent. Robert's art career was centered in Manhattan, while using Merrill Lynch in Santa Fe as his bank.

Thomas owned ten acres of prime land off Tano Road and wanted it subdivided and equipped with water and utilities. As his agent, I learned about subdividing and infrastructure development. I had the land surveyed and subdivided into three lots, a water well drilled, and gas and electricity brought in. By the end of the summer, the development was complete. Thomas had his subdivided land and I had the experience.

In Robert's case, he needed financing to complete the proposed purchase of the studio/home in New York City. Once he had title to the property, he would be able to mortgage it and secure the funds to be the titled owner. I was able to help him, since I also used Merrill Lynch. Using my assets as collateral, Robert obtained the down payment from Merrill Lynch. Subsequently, he was able to mortgage the property and become the owner. Once the mortgage was obtained, my collateral was freed.

Fathers do come in handy and I was pleased to help these two fine sons.

In March Susan and her family, along with Pamela and me, spent a week on a beach we had recently discovered at Puerto Peñasco, Mexico, south of Tucson. In August I joined Susan and Christy, along with his Christy's family, who were being honored at the one-hundredth anniversary of the founding of Estes Park, Colorado. Christy's great uncle essentially founded the city, built the famous Stanley Hotel, brought electric power to the community, and established the Estes Park Bank. (This great-uncle, whose name was Frelan Oscar Stanley, and his twin brother, Francis Edgar Stanley—Christy's great grandfather—were the inventors of the famous Stanley Steamer automobile.)

My last adventure of the year was a tour in November with the Palace Guard to Tucson, and Casas Grandes and Mata Ortiz in Chihuahua, Mexico. At Mata Ortiz, we visited the master potter, Juan Quezada. I was able to do my Christmas shopping at his outdoor studio.

2004: Letters from Caroline

From the start of the year until May 17, I was involved in negotiations with the U.S. Forest Service. The Pecos Ranger District was planning to enlarge the public camping area on Cow Creek, adjacent to my property. The plan involved increasing the overnight facilities, which included space for trailers and automobile parking. The available camping area was already overcrowded and over-used and had worn thin much of the natural charm of the area.

The Forest Service had developed several alternative plans and, as required, invited public reaction. The interested parties, essentially adjacent property owners and outdoor sportsmen, reacted adversely to the proposal. We felt the Forest Service had poorly managed the existing area and that increasing usage with continued sloppy supervision would simply add to the degradation of the area.

With the help of the Santa Fe Conservation Trust and the cooperation of Ranger Reddon and his staff, we developed a compromise solution. The Forest Service agreed to repair, refurbish and supervise the existing camping and fishing areas. If, after a two-year trial period, further expansion seemed warranted, then enlargement of camping facilities could be considered. (As of this writing in 2010, the plans remain dormant.)

I finally completed another activity I had been working on for some time, the publication of *Letters From Caroline,* a compilation of letters my sister Caroline wrote while serving in the Red Cross in the Pacific during World War II. I had long felt I owed Caroline a favor, as I had often been impatient and brusque with her when she was alive. The manuscript had lain dormant in its original typewritten form since 1973. Now, forty-one years later, I had fifty copies printed and published attractively in hardback by the Sungrazer Press of Iowa City. Most of the books were distributed to relatives and Caroline's friends. I felt I had made peace with Caroline with this effort. It completed a family trilogy: *The Buffalo Head* by Daniel T. Kelly, *Dancing Diplomats* by Hank and Dot Kelly, and now, *Letters From Caroline.*

In March I took a trip to satisfy my curiosity about some places in Texas. The state and its people had always intrigued me. Traditionally, New Mexicans disliked *Tejanos.* My feelings were otherwise. I was interested in the minimalist art by Donald Judd and Dan Flavin at the Chinati Foundation in Marfa. I also wanted to see the Davis Mountains and Big Bend National Park. With Joyce Peters as our cultural guide, Susan and Christy and I began our expedition.

Driving from Carlsbad, we entered the foreign land and proceeded to Pecos and then Fort Davis. Our experience at Fort Davis was epic. Located at five thousand feet elevation in the Davis mountains, this re-created frontier post evokes the spirit of post-Civil War military life. It had been garrisoned by Buffalo Soldiers of the Ninth Cavalry. From here, we headed for Marfa, once an isolated cow town and now an illustrious art colony.

Marfa did not impress me. Perhaps my appreciation for art was not sufficiently sophisticated. Or perhaps it was our living quarters. Our hostess was Joyce's friend Gail Gordon, a liberal feminist and faculty member at the University of Texas, Austin. Gail lived in Austin but shared a second home in Marfa with a male friend; she extended her hospitality to the four of us in a house that would have been more comfortable for two. My bedroom was the transit way to the only bathroom.

The Chinati Foundation and the Dan Flavin and Donald Judd art exhibits were located in the converted artillery sheds of Fort D. A. Russell, a World War II-era relic. Judd's creations are superbly crafted aluminum boxes that fill several contiguous brick buildings. Dan Flavin's exhibit is a series of neon lights casting their glow and shadows along the storage shed walls. These creations simply did not capture my imagination.

Leaving Marfa, we drove through a rugged desert canyon down to the village of Rociada on the Rio Grande, located on the thicketed border between the United States and Mexico. From there, we headed for the Chisos Mountain Lodge, a resort tucked into a bowl in the Chisos Mountains. We spent the next day along the towering cliffs of Big Bend National Park, picnicking on the bank of the Rio Grande at the mouth of rugged Santa Elena Canyon. At this isolated spot the mighty Rio was a trickle. Christy and I became "reverse wetbacks;" we crossed the river into Mexico and quickly returned.

Our return to Santa Fe led us through a string of typical West Texas towns—Marathon, Ft. Stockton, Monahans, Kermit—and into New Mexico at Jal. We stopped at the San Simon Ranch, operated for over one hundred years by Joyce Peters's family, then continued on to Carlsbad, where we left Joyce, and then to Santa Fe. My one regret about this interesting trip was not seeing Midland, Texas, George Bush's hometown.

On returning from the Texas trip, the Ides of March set in. Typically, the advent of spring introduced unpleasant windy, dusty and cold weather. I managed an escape, joining Wooly Henry and John Scanlan on a hiking tour in the Amalfi area of Italy. The three of us were members of the Santa Fe Chili and Marching Society and enjoyed hiking together. I had received a flyer from the Wayfarers, a hiking tour organization, extolling the charm of an adventure in Italy. We signed on for the trip for early May. John Scanlan preceded us with a week's hike in Sicily and joined our group in Naples.

On our way to the Amalfi coast, Wooly and I flew to New York and then to Rome, headquartering in Hotel d'Angleterra. We spent five days exploring Rome, taking a private tour of the Vatican, which included viewing the recently restored Sistine Chapel. In the Vatican museum, we were amused to note that the private parts of figures in the nude statuary, most of which dated from the Roman imperial period, were discretely shielded by fig leaves. During the day, we were avid tourists exploring the Forum and Coliseum, trekking along the Tiber and visiting the famous sites of the Eternal City. Gourmet Wooly chose dining spots, seeking out the best food, wine, and general ambiance. We were slightly frayed and wined out by the time we joined our hiking group in Naples.

Our group of five men and five women was led by Vincenzo Falcone, an attractive young man with a pace that was a challenge for us to maintain. We ventured along the steep coast of an area only about twenty miles long. Our treks typically started from the seashore and followed trails up the steep valleys to the summits of the Littori range and then down adjoining valleys to villages tucked in protective coves of the Tyrrhenian Sea. The easiest and most scenic hike was on Capri, a beautiful island free from automobiles, the site of the imperial palace of Caesar Augustus's son.

Unfortunately, at Positano Wooly's heart went out of synch and began fibrillating. John Scanlan and a fellow hiker, Dr. Dina Green, recognized the problem and took him to a local infirmary where his heart regained its proper rhythm. This was the end of the hiking for Wooly. Fortunately, the seizure took place towards the end of the trek. We spent our last day at the ruins of Pompeii; an eruption from Vesuvius in 79 CE buried the city. It was discovered in 1770. Being there was an awesome experience, particularly with nearby Vesuvius exhaling a plume of smoke into the blue Italian sky.

Our tour ended in Naples. We returned to Rome and took the long flight back to New Mexico. Needless to say, none of us was clamoring for immediate hikes.

In June I had a visit with Jeanne Meier, the daughter of Robert Wise, my wife Jeanne's brother. Jeanne Meier never knew her father, who died of wounds in the Philippines. She became close to her Aunt Jeanne, my wife. I gave her photos of her father and her grandparents and endeavored to furnish her details that would bring these relatives into closer focus for her.

In August I was a guest of John Scanlan at his family reunion at Ted Turner's Vermejo Park Ranch. Fishing, hiking and chasing elk herds made this a memorable experience. I was surprised to see such a cluster of gas producing wells on the ranch, but they were inconspicuous compared to the many grazing elk we spotted on the range.

I made another probe into Texas and its history in November, when I spent five days in the San Antonio area on a tour organized by the New Mexico History Museum's Palace Guard. This part of Texas was the scene of much effort by the Spanish to buffer the frontier from French penetration coming from New Orleans. In 1836 it was a battle zone between Mexico and the Texas revolutionaries in the Texans' successful attempt at independence. During the Civil War, the unsuccessful Texas invasion of New Mexico was initiated from San Antonio. The city's museums, galleries, architecture, and Venetian adaptation of the San Antonio River gave the city an unsuspected charm.

On December 18 my brother Booker gave up his long struggle against a contrary heart and died in San Luis Obispo, California. We held a memorial celebration in Santa Fe at Cristo Rey Church, followed by a reception at La Fonda on February 26.

On Christmas Eve, Pamela, her friend Bill, Bob and I froze while watching the Nambé Pueblo nighttime dances. Christmas dinner at my house was attended by family and friends.

2005: A Living Treasure

Fortunately, health problems were more imaginary than real for me at the age of eighty-four. A little shortness of breath on hikes, occasional dizziness, myopic vision—my imagination distorted these minor symptoms into serious problems. In January and February I underwent several examinations and heart tests prescribed by my doctor. I think he ordered the tests just to assuage my worries. In March a review by Dr. Wyndham of the Heart Institute put my worries about my heart to rest. He found no problems and reassured me that I need not worry about its function.

Of more real concern was a persistent aggravation of my right hand. Hot flashes and unpleasant numbness in it disturbed my sleep. This condition was diagnosed as carpal tunnel syndrome, and a surgical procedure in August eliminated the problem. I felt very fortunate that I could continue hiking, playing squash, and enjoying the exertions of rehabilitating the Cow Creek property.

With warm weather finally arriving, I launched into house maintenance, partly just to keep up with the high standard set by my neighbor, Jim Baker, whose recently remodeled home highlighted the worn condition of mine. From early May and on into June, the house was re-stuccoed and the worn brick coping repaired. I lived with a three-story web of scaffolding and the noisy music of the Chihuahuan plaster crew that swarmed over the building. Determining the correct color coating was torture. I delegated the decision to Pamela, Susan, and Juan Aguirre, the contractor, who selected a suitable adobe coloration. The job was done well and the house could look at its neighbor without flinching.

While the noisy home repairs were in session, I sought the peace and tranquility of my Cow Creek sanctuary. The logging roads we built in 2000 were now worn and overgrown. It was time to repair them. I engaged Eloy Gonzales of Pecos to do the job. Eloy was a fixture in the area. A large man, handsome and dignified and around seventy years of age, he was a virtuoso with earthmoving equipment. He and his road grader operator, Frank Encinias, performed cosmetic surgery on the road and also enlarged the stock tank in the center of the property.

Big events were stirring on the world scene. In April a new Pope was elected, Benedict XVI, a German theologian. He succeeded the globe-trotting Polish pope, John Paul II. A terrorist bombing in the Metro line rocked London in July, and in August, tropical storm Katrina ravaged New Orleans and the Gulf Coast.

John Agresto, former president of St. John's College, began an effort to establish a western college in Iraq. On October 28 Michael Peters was inducted as the sixth president of the St. John's Santa Fe campus. On November 6 Sam Adelo, Viola Padilla and I were recognized as "Living Treasures" of Santa Fe. My feeling at the time was that my qualification was simply that I was living, while the others were the real treasures.

In March I spent four days in the Carlsbad area as a guest of Joyce Peters. That southeastern part of New Mexico was somewhat removed from colonial Spanish and Mexican influence. It was settled by Anglo ranchers, farmers, and venture capitalists at a time when the Apaches were trying to hang on to their nomadic lifestyle. During my visit, we hiked the Guadalupe

Mountains and Capitan Reef and probed Carlsbad Caverns. I was fascinated by a mescal ceremony performed by the nearby Mescalero Apaches.

I organized a family reunion in early April. The children and their spouses and offspring gathered at Los Caballeros, a ranch-resort near Wickenburg, Arizona. For five days we visited and became reacquainted. We welcomed the springtime weather in Arizona, such a contrast to dusty, windy New Mexico. The desert was blooming and the weather balmy. It was a pleasant reunion.

In June I joined a tour organized by the Museum of New Mexico Foundation and the Palace Guards, a four-day excursion starting in the Belen-Tomé area south of Albuquerque and then moving on to Zuni Pueblo and into Arizona. We took in the Petrified Forest National Park and the Hopi Indian reservation, and spent the last night in Winslow at the Posada Hotel, a relict of the Harvey hotel system with its famous Harvey Girl waitresses and hostesses of the railroad tourism era.

I was particularly interested in the Belen-Tomé area. Located in the Rio Grande Valley, Tomé was once a prosperous agricultural oasis along the route of the Camino Real from Mexico to Santa Fe. Tomé Hill overlooking the village was a welcomed landmark for the ox trains delivering supplies to the settlers of this remote area. During my Gross Kelly days in the 1950s, Tomé was the headquarters of a thieving ring where goods stolen from our Albuquerque warehouse were fenced. I helped the state police to catch the operators, who were later convicted.

In September I visited with a number of my friends in Denver. My college roommate Joe Downer was there visiting his daughter—who is my goddaughter—Jeanne. Joe was in poor health, and I was anxious to see him. Jeanne gave a dinner party and guests included Jack Malo, David Wilhelm, and Rosie Stewart. I was able to re-cement my bonds with these old friends.

As a special treat, after the dinner reunion I drove southeast 150 miles to Bent's Old Fort National Historic Site on the Arkansas River. The fort was established in 1833 as a trading site on the border between Mexico and the United States. The brothers Charles and William Bent, along with Ceran St. Vrain, developed an important trading enterprise at the fort, which also served as a resting and re-fitting point for the wagon trains plying the Santa Fe Trail.

Of particular interest to me, this nonmilitary fort was also a spy nest for the United States in 1845 and 1846. The Army considered using the fort as the launching point for invading northern Mexico, when or if the war between the two countries took place. In those years, Lt. James W. Abert of the Army Topographical Engineers was developing critical intelligence for the army. He also led a reconnaissance from there into what is now Colorado, New Mexico, Texas, and Oklahoma. His reports later aided General Kearney during his invasion into Mexican territory. Once he had claimed New Mexico for the United States, General Kearney appointed Charles Bent governor of the newly acquired territory. Less than a year later, rebellious Taos Indians murdered Bent. A visit to Bent's Old Fort is a must for those interested in the commercial history of the Southwest.

My final adventure of the year was spending Christmas with Robert in New York City. As usual, I stayed at the Harvard Club. While Robert, the busy artist, was working, I explored Manhattan Island. When he was free we toured the museums and frequented his favorite cafés, accompanied by Shirine, his attractive Iranian companion.

Christmas dinner was special. We were guests of Shirine's sister and brother-in-law, Simine and Purvis Moozami. Following a hybrid American-Iranian dinner (turkey, pomegranate and succulent Iranian rice), we were treated to a pictorial review of political and diplomatic events involving the sisters, whose father, Jafar Sharif-Emami, had been prime minister of Iran under Mohammad Reza Pahlavi, the last shah of Iran. The glamour, splendor, and glitter of that former life was quite a contrast from the quiet family dinner we had just completed.

2006: A Fiftieth Milestone

In January 2006 I was occupied with my the usual activities—Chili Club, trips to Cow Creek, and First National Bank board meetings. The bank leadership shifted from Ed Bennett to Gregory Ellena, who took the bank in a vigorous, new direction. He sought to expand the bank and bring new blood into management and the board of directors. I knew my long tour was soon to end. After thirty years of serving as a board member, I resigned later in the year, on October 16.

In February and March, mild winter weather permitted some outside adventuring. I took a trip with Jim Gordon to explore his Canadian River Canyon ranch, which encompassed over forty sections of land around the rugged gooseneck bends of the Canadian River east of Springer, New Mexico. Much of the acreage was inaccessible, except by helicopter, because of the canyon's steep walls. Early inhabitants of the canyon left evidence of their presence in caves and rock art on the canyon walls. The Spanish names of landmarks in the area—Huérfano (orphan) or Corazón (heart)—reflected the remoteness of the area. The neighboring Bell Ranch sprawled over a huge area. The whole region had a wild, uninhabited character. It was strictly God's country, undefiled by modern man.

I had a bit of an adventure at the Donnelly Library Archives of Highlands University in Las Vegas as I perused photocopies of the *Las Vegas Daily Optic* dating as far back as the 1870s. I spent several days reviewing the papers, looking for information about Grandfather Harry Kelly, who as a young man became an active member of the local volunteer fire department. He was a bachelor and his roommate was young Miguel Antonio "Gillie" Otero, who would later become the last territorial Governor of New Mexico.

My twin sons were to turn fifty on April 16, an epic event that I wanted to recognize. I suggested we take a trip together and left the destination up to them. They chose the Croatian coast and a birthday celebration in Venice, Italy. On April 7 Thomas, who had flown in from Kathmandu, met us at the Venice airport. Robert served as driver in a rental car as we proceeded directly from the airport to Trieste on the Italian-Slovenian border.

From then on, we followed the Croatian coastline, heading south along the jagged, mountainous edge of the Adriatic Sea. Stopping for scenic picnic lunches, we drove from Trieste to Zadar and on to Split. I was left at the Park Hotel while they enjoyed the company of two Ukrainian beauties they had met at the hotel. From Split, we took the ferry to Korĉula, a scenic island, and returned to the mainland, and on to Dubrovnik.

Dubrovnik is a gem. It is encircled by a fortress-like wall that rises from the harbor on an inlet of the Adriatic. No cars are allowed in the port section of the city. From the mountain that looks down into the center of the city, the Bosnian Serbs were able to shell the entire city during the civil war. The bright red tile roofs of the repaired buildings marked the numerous targets that the Bosnians had hit.

From Dubrovnik we headed south to Bosnia-Herzegovina and terminated our adventure in Montenegro, where we were amused seeing the border guards wearing white medical gowns. Their duty was to spray cars to prevent the import of agricultural diseases into this backward, emerging nation.

To be in Venice on the boys' birthday, we needed to cross the mountains from coastal Croatia into the fertile rolling hills of the eastern part of the country. We sped north on a super highway to arrive in Venice on April 15. The Vaporetto, a water bus operating on the Grand Canal, delivered us to Hotel Manin.

Coincidentally, Sunday, April 16, the boys' fiftieth birthday, was Easter Sunday. It was a gloriously bright day. The cardinal primate of Venice was celebrating High Mass at San Marcos Cathedral; we attended the ceremony. The church was packed and the organ blasting. The Cardinal spoke in Italian, French, Spanish and English, wishing the congregation best wishes and blessings on the feast day. It was a magnificent recognition of Robert and Thomas's fiftieth birthday.

Unfortunately, I suffered some injuries during the trip. While toting baggage in Venice, I twisted my right knee, and in Croatia I slipped off a raised deck and tore the rotator cup in my right shoulder. Also, for some reason—probably old age—my lower eyelid collapsed, exposing a bloodshot left eyeball. I was a real mess.

After our celebration, it was time to return to reality. Thomas departed for Nepal while Robert went back to his New York studio and I limped home to Santa Fe, in need of a physical overhaul. For the next month I was in for repairs.

By Memorial Day I was reasonably recovered and decided to call on my deceased veteran friends at the Santa Fe National Cemetery. After paying respect to those resting in the newer, landscaped section of the cemetery, I ventured into the older, southern section, a historical treasure trove. There were unknown soldiers' gravestones, graves of the Confederate dead from the battle of Glorieta and of members of the Seventh and Ninth Cavalries, the famous Buffalo Soldiers. The highlight for me was the grave of Governor Charles Bent, who had been the victim of the bloody uprising in Taos in 1847.

The children own the house I live in at 535 Palace Avenue. I am the tenant, but I assume the cost of some of the renovations. In 2006, these included substituting old toilets with new, replacing the ancient converted coal boiler with a modern, more efficient heating system and stabilizing the foundation of the house by driving several braces deep in the ground.

To escape the chaos of the house repairs, scheduled for early September, I suggested to Wooly Henry we take a trip to the Pacific Northwest. He readily accepted and we flew to Boise, Idaho, rented a car and took off to make a circuit through Oregon and Washington. We attended the famous Pendleton rodeo and then drove to Portland and joined my daughter Pamela and her beau, Bill Wanker, a Portlander. We saw where the mighty Columbia River empties wildly into the Pacific. Not far from there is the site of the winter camp of the Lewis and Clark expedition, later named Fort Catsop. What a cold and wild winter that must have been for those intrepid explorers. Leaving the Portland area, we visited the former fishing complex of Astoria, now a victim of tourism exploitation. We traveled north as far as Port Angeles on the Strait of Juan de Fuca in Washington, then returned to Portland for a succulent dinner of Dungeness crabs prepared by Bill Wanker's parents, Pauline and William.

By the time we returned to Santa Fe, the repairs on the house had been completed. I paid the bills and moved back into a renovated home.

I spent October thrashing with indecision about purchasing a unit at Quail Run. The house was part of the estate owned by my cousin, Patricia Parker, who had died there during the summer. I had considered having a hip pocket home in the event I needed a simpler lifestyle as time went on. Houses at Quail Run were in considerable demand and Patricia's son, Tom Parker, offered the house to me. Should I buy or not? I inflicted my indecision on my children and friends for some time. I finally decided to buy the home and on October 31 became its owner. It has been a rental unit since then.

In November I was busy supervising the fencing of the children's Tano Road property. A substantial development was in progress along the western boundary, which had never been fenced. I engaged the Valencias, my veteran Pecos crew, and with Bill Wanker's help spent several weekends erecting a fine, four-strand fence along this border. It was tough work, but pleasant. The finished fence reminded me of my old cowboy-ranching work days.

The Christmas season was classic. During Christmas week, Santa Fe was blanketed with over two feet of snow, and skiers were happy. The town was blessed. I hosted sixteen people for Christmas dinner after viewing the Cloud Dance at Tesuque.

2007: Venture Capitalism and a Pilgrimage

The economy was vibrant. The stock market was at a high. Credit was obtainable. Speculation was almost riskless. This national optimism spread to my family. I had become the owner of a house at Quail Run and was anticipating receiving a substantial monthly rental income to offset the purchase price. I felt quite secure with my investment portfolio and had an accommodating attitude towards assisting my children in their financial pursuits.

Susan and Christy, joint realtors buttressed with Christy's contractor's license, were interested in purchasing, remodeling and selling an attractive prospect on Gildersleeve street in Santa Fe. I helped them with the purchase down payment. In September the renovation was complete and the house quickly and profitably sold.

Pamela was having considerable success in luring manufacturers and retailers into developing products modeled after items from the Museum of New Mexico collections. Her beau Bill had obtained a hi-tech patent with the potential for being highly valuable. While he was marketing the concept, he was designing computer programs for the New Mexico Public Education Department. He and Pamela both felt comfortable enough financially to take a two-week horseback trek in Patagonian Argentina in honor of Bill's fiftieth birthday.

Thomas was favored with an ambitious showing of his work at the Verve Gallery in June. He spent a month as my housemate as he prepared for it. Concurrently, he was organizing a challenging international extravaganza commissioned by Gregory Colbert to be filmed in Mongolia later in the summer. This was to be part of Colbert's "Ashes and Snow" project, an international photographic traveling museum. Typical of Colbert's passion, this project was an unprecedented film concept interpreting the relationship between man and animal, ultimately to be shown worldwide. Thomas, a Mongolian wilderness expert, was assigned the task of staging the project. This required bringing reindeer, captive eagles, camels and horses into an appropriate setting for filming. It also required casting the Mongolian herdsmen needed to orchestrate the animal and human relationships involved.

To my amazement, Thomas was able to initiate this complex assemblage using my rudimentary telephones and computer. Later in the summer, he photographed the melodrama in a remote section of Mongolia. It was highly successful and Thomas was compensated adequately by Colbert. The project has since been exhibited internationally. Thomas continues his adventures in Mongolia during the summer.

The artist Robert was also having a successful year. In 2005 he had a show in Italy that had developed into a close artistic relationship with his Milano art dealer. As a result, he became familiar with the Italian art scene, particularly with the works of deceased artists. Feeling comfortable about the

2007 art market, he held successful shows in Aspen, Santa Fe, and Monterrey, Mexico. During this period, he became interested in speculating in art and acquired a painting by the deceased Italian artist, Mimmo Rotella. Robert's Italian dealer suggested he attempt to acquire the painting at an upcoming auction, anticipating the painting would certainly sell at a premium in the near future. Intrigued, Robert obtained a fellow investor and, because he needed funds, impressed me with the potential value of the speculation. For a three-month period, I made available a loan. Robert and his partner in the investment bought the painting. By adjusting the mortgage on his New York studio, he obtained the funds to repay me for my loan in late June. My timing was propitious.

My actual investments were in adventure capitalism, rather than venture capitalism. My first was a three-day visit to San Francisco at the invitation of Pamela and Bill to celebrate my eighty-sixth birthday on March 22. We hiked to Stinson Beach, took in a ballet, watched part of the California State Doubles Squash tournament, and spent too much money on refurbishing my wardrobe at Nordstrom's.

In April I was John Scanlan's roommate while we, along with Beth and Wooly Henry, explored the lunar landscape in the Valley of the Gods in southeast Utah. Our journey took us through the Navajo Nation to Blanding and the Mexican Hat region along the San Juan River. The Valley of the Gods has not been spoiled, unlike its neighbor Monument Valley. This area is virtually unoccupied. We stayed at an isolated bed-and-breakfast inn located at the foot of towering Cedar Mesa, whose summit is reached by a harrowing route chiseled from the mesa's cliff face.

Upon our return, Beth rescued an abandoned Navajo puppy whose name is Moqui Dugway, in honor of the Cedar Mesa roadway. The dog is now as fat as many of the fast-food-eating Navajos who abandoned him.

Before the August stock market collapse, I organized a trip to Spain. In purgation of my sins, I felt the need for some comfortable penance. The plan was to be a *peregrino,* or pilgrim, on El Camino de Santiago, but we planned to compensate for the footsore suffering by indulging in creature comforts at comfortable stops along the route.

In September the sincere pilgrims, Pamela, Bill and I, were joined by the hedonists, John Scanlan, his friend Nola, and Ramon, the driver of a fully equipped eight-passenger limousine. Our journey began in Barcelona, where we spent three days, somewhat overwhelmed by the architectural triumphs of

Antoni Gaudí. From there, we flew north to San Sebastian, where we met Ramon and his limousine. We visited briefly in France at Saint Jean de Luz and Biarritz, then spent a day in Balboa at the Guggenheim Museum, again influenced by an architect, Frank Gehry.

We went south and west and walked the pilgrim's trail for a series of short day hikes of six to seven miles. At designated points en route, we were met by the sinful ones in their limousine and were taken to non-pilgrim *paradores* for rest and rehabilitation. The hikes were stimulating and somewhat spiritual. We enjoyed meeting the fellow pilgrims and felt a part of the mystery and history of the pilgrimage. Our last day we hiked into Santiago to the pilgrim center, where we had our *Credencial de Peregrino* officially stamped. Like the real pilgrims, we wore our Peregrino scallop shells with fraternal and humble pride.

From Santiago we toured and detoured to Salamanca, Avila, Segovia and into Madrid, where we ended the trip. One exceptional detour we took was to a bull ranch, the Ganadería Herederos de Miguel Zaballos, some forty miles south of Salamanca. The ranch raised the black fighting bulls of the *corridas*. The breeding herd consisted of some one-hundred mother cows especially selected for their proportions, horns, and combative dispositions. The bull calves chosen for fighting are separated from their mothers as yearlings. Each year thereafter, they are again separated by age and build. After three years, they are inspected by buyers for the ring. The bulls are never teased or enraged by the growers; their fighting capabilities are evoked by the toreadors in the arena. A fighting bull may command as much as forty thousand dollars. Bull fighting is still popular but its acceptance is shrinking as distaste for animal cruelty increases.

In Madrid, Scanlan and I again took most of the walking tours of the city. My last day before returning home was an all-day visit in the Prado. The royal portraits compounded my confusion with the complicated Spanish monarchy, which historically included the Habsburgs, Bourbons, and others. The monarchy, now a constitutional institution, still plays a role in Spain.

On returning home, my interest in New Mexico history was renewed by reading a great novel, *Blood and Thunder*. I was intrigued about the route the Navajos followed on their forced relocation to the Bosque Redondo on the Pecos River. This then led to curiosity about the geography and history of southeastern New Mexico. It just so happened the Palace Guards of the New Mexico History Museum was offering a tour of southeastern New Mexico in early October. I signed on.

We spent our first day at Fort Sumner, the military garrison monitoring the million-acre Bosque Redondo Indian Reservation, where the Navajos were interned. The route of the Long Walk, as the Navajo's four-hundred-mile forced migration was called, began in the Fort Wingate area near Gallup and led eastward along what is now I-40 to the Bernalillo area. From there, it probably went north along the Rio Grande to Santo Domingo and west to the upper Pecos drainage. The route continued down the Pecos River through San Jose, Ribera, Villanueva, Anton Chico, and Dilia, and on to Fort Sumner and their new homeland.

Following the abandonment of the Bosque reservation, white settlers moved in. Lucian B. Maxwell, an affluent rancher from northwest New Mexico, migrated from the Cimarron area to Fort Sumner, where he again prospered and ultimately died. It was here, in Pete Maxwell's house, that Pat Garrett killed Billy The Kid in 1881.

From Bosque Redondo, our group went to Roswell and Ruidoso via San Patricio, then to the Sentinel Ranch where Peter Hurd created much of his art and where his nephew, Michael Hurd, continues the family's artistic tradition. We visited Lincoln, where we were briefed on the Lincoln County War. Here Billy The Kid earned his fame as a killer and escaped from the jail. We also visited Fort Stanton, where General "Black Jack" Pershing served as a lieutenant.

From the lower Pecos country, we crossed the mountains westward to Trinity Site, the site of the first atomic bomb test on July 16, 1945. The awe-inspiring explosion proved the bomb would work, and just a few weeks later similar devices were dropped on Japan, ending World War II and ushering in the atomic age. The isolation of this wasteland testing area and its place in history impressed me greatly. As a soldier some sixty years earlier, bound for Japan, I cheered President Truman on his decision to drop the devastating bombs on Hiroshima and Nagasaki.

From November on for the rest of the year, the collapse of the economy dominated the news and led me to much reflection and concern. However, life went on and I stayed busy splitting firewood, playing squash, picking and disposing of an ample pear crop and beginning a review of my diaries. The death of two close friends—Newt Eddy and Joe Downer, my college roommate—ushered in December and made me reflect on the preciousness and brevity of life.

Joe Downer's passing was an especially severe emotional blow. I reacted by composing a eulogy for him, based on our long friendship. It was my personal tribute to him.

Pamela and Robert came home for Christmas. Susan and her family were vacationing in Puerto Rico, while Thomas, as usual, was either in Kathmandu or Goa, India. Pamela hosted a Christmas dinner at my house and afterwards we attended Mass and a Tesuque Indian dance.

On the last day of the year, Robert and I drove to Cabezón Peak, the great sentinel of the Rio Puerco area. We hiked to its base and as we viewed the lonesome and rugged country, we reflected on how special our lives had been. The deep reflection passed when I discovered I had lost one of my hearing aids on the rocky flank of Cabezón.

2008: Financial Frustration

This was a sour year for the whole nation, a time of progressively growing fatigue, frustration, economic ruptures, and political turmoil.

The wars in Iraq and Afghanistan seemed to be pointless, internecine conflicts beyond America's power to resolve. The Bush administration had largely polarized the nation. The euphoric economy and the stock market suddenly took a nose dive as the financial engine running the economy exploded. Since this was a presidential election year, the political tempo whirled continuously faster.

All of this had its effect on my tiny sphere of activity. In New Mexico Heather Wilson was running for the Senate against Santa Fe's own Tom Udall. Nationally, Obama and Biden won the Democratic presidential nomination. McCain and Palin were the odd-couple Republican candidates. Although I was a registered Republican, my instincts were to remain independent of rigid party affiliation. I supported Heather Wilson and I felt a certain amount of sympathy for Bush, mostly because of the constant bullying he received from the press and from the far-left liberals. However, Obama stimulated the prospect of a fresh breath of air and youthful "hope," and I voted for him.

The thunderous stock market crash of August hit me hard, along with everyone else. This was particularly so with my Merrill Lynch account. I had been with Merrill for many years and had a close relationship with my advisor. I was shocked when, on September 14, Bank of America bought Merrill Lynch at a bargain price. The knock-out blow came when my advisor resigned and moved to Morgan Stanley. I was suddenly an orphan in the financial world. My case of financial flu was painful; my capital weight was down but I was still in relatively good shape.

As the economy nose dived, I seemed to suffer a sympathetic reaction in my physical state. At eighty-seven, I was noticing some depreciation. In March my dentist, Dr. Spier, suggested I cap my front teeth. I followed his advice and gained a white, toothy smile that apparently improved my looks and my digestion.

Despite the wars, politics, depression and health concerns, the year had some interesting distractions. The annual Hopi Bean Dance took place on February 15. Pamela, Bill and I drove the long haul to the Second Mesa Visitors Center, where we spent the night. Before daybreak, we drove several frigid miles to Hoteville. There, Crow Mother and her green bean sprouts appeared at sunrise and circled the village, dispensing the life-giving sprouts to the villagers who lined her route. At noon, the Kachinas emerged out of the six kivas. Their masks were varied and wonderfully imaginative. It was a fascinating ceremony and seeing it was well worth the effort. During the ceremony, we were lunch guests at the home of Bob Rhodes and Verna Lolama.

In June I attended my sixty-fifth reunion at Harvard. The tolling of the bells and the memorial for the many departed classmates gave a bittersweet flavor to the otherwise pleasant event. I went from Boston to New York to visit Robert, staying in the city for several days to tour the city and the artistic activities around which his life was centered.

In late July the annual National Furniture Mart took place at the enormous convention center in Las Vegas, Nevada. Pamela had several clients of the Museum of New Mexico Foundation displaying at the center. Also, she was to give an address on aspects of New Mexico's Hispanic culture. I decided to go see her in action. I had not been in Vegas in years, so the growth in glitz was staggering. I stayed at the Belagio.

I met Pamela's clients and was impressed by their enthusiasm for their New Mexico-influenced products. I was also impressed by her talk on New Mexico history and its tri-culture heritage. The presentation was well received by a large audience. I was proud of her but not impressed with Las Vegas. In fact I was turned off by all the fakery and greed that Vegas displayed. I had become a prude in my old age. I was glad to be returning to Santa Fe.

In August Jerry Peters invited me to his annual horse-ride and outdoor weekend at Quinlan Ranch. The ranch near Chama is an elk breeding and trophy hunting facility. Some fifty to a hundred outdoorsmen guests gathered at the ranch. Excellent meals were prepared al fresco by the visiting staff of Jerry's Santa Fe Rio Chama Steakhouse. The guests, on their own or rented horses, took guided rides to selected parts of the ranch. In the evening, guitarists and western folk balladeers entertained the guests clustered around a large circular fireplace.

My primary interest was in Jerry's elk breeding operation. A select herd of elk cows range in a 1,200-acre fenced pasture. They are artificially inseminated with semen from trophy bull elks. The resulting male calves are selected for their trophy qualities and, ultimately, are hunted or sold to other operators.

For the final dinner, a butchered buffalo was roasted for several hours in an outdoor pit and served to the salivating guests. Jerry demonstrated he was not only a renowned art dealer but also a strong conservationist, apparent in his care for the ranch and the animals.

On my own rancho at Cow Creek, I spent my weekends on maintenance and improvements. As time went on, standing trees killed by the fire began to fall on the new, mile-long fence we had recently built, snapping the wires. I had to continually patrol the fence line and repair the damage. One of my trusted workers, George Valencia, came up with the idea of taking preventative action by felling all the trees that could fall on the fence. This would eliminate the need for constant patrolling. Following Geroge's suggestion, the crew worked on both sides of the fence and eliminated potential fence breaks along the whole property.

Also replaced was the underground pressure tank that held the water for irrigation and domestic use on the ranch. The original twenty-year-old pressure tank had ruptured. It was housed in an underground pit about eighteen feet deep, accessible through a steel culvert four feet in diameter. It was a tight fit for anyone who had to descend and work on the pressure tank.

George Boylin had put in the original pressure tank. I had not seen him in years, yet I knew he was still a maestro in drilling and outfitting water wells. I called George, who was still grumpy and antisocial. He agreed to inspect and, if needed, install a new system in the existing underground location. He arrived with his helper, his wife Anita. To my surprise and amusement, George had grown as fat as Santa Claus. His girth was greater than the opening to the well housing. In order to descend into the pit, he used a crane mounted on his pickup. The crane hoisted him up and jammed him into the vertical culvert. He descended, diagnosed the problem, and called for the crane operator, his calm and competent wife, to pull him out.

Not only did George renovate the system, he installed two tanks in the same space. Despite his size, he connected all the fittings and modernized the system. When he was done, he declared this was his last job and he was retiring.

As the fall approached, my voluntary free-firewood cutting season commenced. I moved the splitter to the property and, with the usual assortment of reluctant volunteers, supplied the winter's wood needs to the Kiva Club and a number of friends, both needy and otherwise.

This turbulent year ended with my interest centered on two new projects. One involved genealogy and the other was a diary review of the some sixty years of entries I had accumulated. I was looking forward to a busy new year.

2009: A Reminiscent Phone Call

Thomas's Nepal family, who had spent Christmas with Carroll's father in North Carolina, descended on Santa Fe and into my house at 535 Palace Avenue for a weeklong visit on January 2. I moved out of the house to a suite at La Fonda Hotel, leaving the house to the family—Thomas, Carroll, Liam (thirteen) and Gaelan (nine). I moved for two reasons: to allow them to enjoy the house privately and to see if they would want to move in permanently; and to avoid the chaos that would undoubtedly reign when the boys and their Santa Fe cousins joined forces. The plan was only partially successful. After two days at La Fonda, I moved back. I disliked hotel living and missed the family action.

The visit was a success. With their departure, my solo life settled back into its normal routine, disrupted only by the gyrations of the economy and, especially, of the volatile stock market, which cratered in January and February.

Somewhat tired from worrying about my portfolio, I accepted an invitation to visit my old Navy friend, Victor Delano, in Naples, Florida. My birthday was coming up, and I never relish spending March in Santa Fe. A few days in comfortable and somewhat depression-proof Naples sounded fine. From there, I planned to visit Robert in New York to celebrate my eighty-eighth birthday on March 22.

At balmy Naples I enjoyed the beach action. Along with watching the bathing beauties and observing the fishermen along the extended pier, Victor and I shared some excellent meals. His apartment was comfortable. Listening to his naval recollections and taking late afternoon naps calmed me. The only drawback to the visit was my inability to complete the crossword puzzle that was the daily challenge at breakfast time.

From Naples, I flew the coastline north to New York, arriving on March 21. Robert met me, and my room was waiting for me at the Harvard Club. After a great steak dinner at Wollensky Steakhouse and some single malt scotch, Robert returned me to the Club. The next day, my birthday, we attended Mass at the quiet and charming neighborhood church of St. Thomas Moore, where Jeanne and I had married fifty-six years before. It was a fitting and sentimental way to start my new year.

Having been impressed by Victor Delano's naval stories, I suggested we board the aircraft carrier *Intrepid*, which was permanently moored at the 45th Street Pier as a naval museum. The carrier was awesome. Its wartime crew numbered over three-thousand sailors. It had births for over one-hundred fighter planes. The ship had been actively engaged against the Japanese navy during World War II. My lasting reaction to that visit was pondering on the tremendous responsibility carried by the captain commanding such a vessel.

Revitalized by my New York visit, I returned to Santa Fe. Springtime projects included repairs to the house and continual landscaping and aspen clearing on the Cow Creek property. In June I commenced an ambitious project: reviewing a lifetime's worth of diaries. My father kept diaries from 1916 until his death in 1974, while my diaries began in 1942 and were ongoing. Somewhat pushed by the fact that three of the family had published books (*The Buffalo Head, Dancing Diplomats* and *Letters From Caroline*), I felt a responsibility to pursue a similar effort.

In order to compose a chronology for 1921-1942, when I didn't keep a diary—I needed to review my father's notes. Fortunately, I was familiar with many of his entries because they involved activities of our family business, Gross Kelly & Co. Thus, my task began with my father's comments on the labor strife in Colorado in 1916 and covered both business and family entries involving me from my birth until his death in 1974. This project, along with other distractions, took up considerable time for the balance of 2008. I ended the year having reviewed the diaries as far as 1955.

On July 21 Florentino Valencia, Jr., the uncle of my Cow Creek foreman Gary, died in Pecos and the age of eighty-eight. I attended the funeral. The Valencia family was prominent among the old family Hispanics of the

Pecos valley. After the funeral, I met some of his kin. I was entranced by some of their names. Florentino's deceased foster parents were named Tiburcio and Lucianita Roybal. His biological parents were Florentino and Crisostoma Valencia. His several brothers and sister had carried the names Miguel, Pablo, Mariano, Fermín, Mercy, Liborio, Fedelina, Refrugito, and Erinea. These old names, along with their bearers were, indeed, endangered species.

In July it was my turn to write a paper for the Chili Club. I had selected as a topic Harvard's recent president, the controversial Lawrence Summers. Coincidentally, that month, *Vanity Fair* published a detailed article, "Rich Harvard – Poor Harvard." The college had suffered severe losses in its endowment portfolio, and much of the loss was attributed to actions taken by Summers while he was Harvard's president, when the endowment shrank from $38 billion to approximately $26 billion. My Chili Club paper was timely. It did not, however, solicit much sympathy from my non-Harvard listeners.

In September I was reviewing my 1952 diary and reading an entry about an enjoyable, romantic horse ride I took with Laura Maud MacArthur (her married name is Peterson) on her ranch in Wagon Mound. Coincidentally, the phone rang, I picked it up and was startled by the voice on the other end. "This is Laura Maud MacArthur." It was the voice of the rider I was with fifty-seven years ago, whom I was just reading about. By now, she was widowed, eighty-three years old, and residing in a retirement home in Charlottesville, Virginia. Shades of mental telepathy. Homesick for New Mexico and evidently warmly remembering me, she wanted to reestablish our relationship and move back to Santa Fe. (As I write this now, our timeworn relationship continues via the U.S. Mail.)

By 2009 my daughter Pamela was in her eighth year as director of licensing and product development for the New Mexico Museum Foundation. Anxious for her success and interested in supporting the foundation's aid to the state museums, I had been making an annual contribution, which put me on the foundation's mailing list. This October the foundation was offering a tour of Boston. Although I had spent six years at Harvard, there was much of Boston I had never seen. The thought of a guided trip there intrigued me and I decided to go.

From October 19 to 23, I fell in love with Boston all over again. The tour was organized by the Berks, former owners of the renowned Berklee College of Music in Boston, now living in Santa Fe. Among the interesting places we visited were the Massachusetts Historical Society, Berklee College of Music, JFK's Library and Museum, the Museum of Fine Arts, the renowned Gardner Museum, and Beacon Hill residences. Our final dinner was held at

Harvard's Faculty Club, where we were entertained by my friend Dwight Miller, the Harvard admissions recruiter for New Mexico. I was impressed by how far Boston had come since the trying years of World War II; it was once again a charming and cultivated city.

In contrast to my visit in sedate Boston, putting the Cow Creek rancho to bed for the winter resembled a scene from the Wild West. In October and November local grazing permittees move their cattle from their allotments high in the mountains to the winter pastures around Pecos. The Valencias had gathered their cows and were moving them down the Cow Creek road, adjacent to my property. Suddenly, one of the cows was frightened and bolted from the herd in panic. She jumped over the boundary fence and disappeared into the aspen grove within my pasture. Unaware of the escape, the Valencias continued the drive on to Pecos.

I was not aware of the cow's escape, either, but when I visited the rancho a day later, I was surprised to see a black cow running into the shelter of a thick stand of aspens. The lone cow was still wild with fright. When I returned to Pecos, I mentioned the incident to Gary Valencia. He immediately deduced the wild cow was the one missing from the family's herd. Efforts to rope or corner the crazed cow failed and the animal evaded its owners, jumping the fence again to head back up the road towards the mountains where it had summered. The animal was never captured and was finally shot and butchered where it fell.

In late November my old friend Governor Bruce King died. I knew him from my ranching days in the Estancia Valley. I went to Stanley and Moriarty to attend the mammoth memorial, where former President Bill Clinton delivered an eloquent tribute to Bruce. Those two characters—King and Clinton—were consummate and very successful politicians.

During Christmas week, I hosted a large party at La Fonda, with the Kathmandu Kellys as the featured guests. It was my way to reciprocate for kindnesses rendered me by friends and to reintroduce the Thomas Kelly family to their many friends in Santa Fe.

2010: Looking Back

The advent of this new decade marked the sixth month of work on my autobiographical diary summaries. During the year just ended, I reviewed my father's diaries from 1916 until 1960. Combining his recollections and my own notes, I was now up to 1959. It had been a time consuming experience. My objective now was to complete the project by the end of 2010.

My romance with Laura Maud MacArthur was becoming more serious. It was a somewhat intriguing concern. Her letters came almost daily. Her fantasy blossomed into an escape melodrama. I was to free her from Virginia and her controlling daughter, take her to New Mexico and marry her, thus fulfilling her lifelong desire to unite with me as her lover.

I became concerned her fantasy might become a reality. I did not want to hurt Laura Maud in any way. Yet, I felt it was time to set her straight. I wrote a letter to her and sent it to her daughter Laurie for her review. In it, I acknowledged our long but disjointed friendship and expressed my desire it would continue to thrive. However, I emphasized my status as a longtime widower/bachelor. I explained I was settled in this role and although I was honored by her affection, I did not wish to remarry—and in any case I considered myself, at eighty-nine years of age, to be ineligible.

Laurie approved of the letter. She felt Laura Maud might be hurt at first, but her memory would soon be salved over and no emotional harm would result. When Laura Maud received the letter, she stopped writing to me for several days, then began again. Now the thrust of her letters was not matrimonial but rather expressed a desire to visit her beloved New Mexico. I felt I had cleared the air and the immediate challenge to my bachelor life had been averted, but a visit was brewing.

I felt I deserved a little relief and relaxation. I promoted a stag trip to warm Tucson. John Scanlan and Wooly Henry, companions on the hike in Amalfi, Italy, also felt justified in getting out of town. We drove to Tucson and stayed at the comfortable Arizona Inn. Here we were joined by two friends from California, Sandy Powers and Phil Hudner.

This congenial group spent March 1-5 visiting different sites in the area. Our tour took us to the Arizona Desert Museum and the Elkhorn Ranch near Sasebo, a guest ranch and game preserve of some twenty-seven thousand acres. We had a private tour of the mission church of San Xavier del Bac, built in 1692. We spent an afternoon at the Pima Airplane Museum. Our last day we visited the University of Arizona campus and the ski resort on Mt. Lemmon (which my agency used to insure). The ski area is a scant two to three crow-flight miles from Tucson and boasts skiable snow at its nine-thousand-foot elevation.

The only drawback to the Tuscon trip was the poor restaurant choices made by our California friends. They insisted on eating at two Tex-Mex chile parlors. We New Mexicans knew our real chile but we tactfully went along with our California friends' choices.

In April the British Petroleum well in the Gulf of Mexico blew up. This epic blowout absorbed the nation's attention, but here in Santa Fe life went on in its usual rhythm. Pamela was accepted at Christie's Education in London to pursue a Masters in Renaissance Art and Culture. Her year course was to begin in September. Knowing this would entail a long separation from me, she began promoting a father-daughter adventure.

This idea coincided with a resumption of letters from my Virginia connection, Laura Maud. She had recovered from my matrimonial rejection but continued reminiscing about New Mexico and her desire to return. By now Pamela was feeling empathy with Laura and suggested we include a visit with her in Charlottesville on a tour of historic places in Virginia. This appealed to me.

Pamela and I spent a delightful time together May 13 -17. We headquartered in Richmond, where I sought out information on where my great-grandfather had lived and something about his medical practice in Richmond during the Civil War. I found the street where he lived and the former location of the hospital he served. We also visited the American Civil War Center and the White House of the Confederacy. The Civil War was still alive in Richmond.

From Richmond, we drove to Yorktown and toured the battlefield where Cornwallis surrendered British forces to Washington and Rochambeau. The British literally had their backs to the sea, and the blockading French navy denied them from being resupplied—thus, the epic surrender.

En route to Charlottesville and Laura Maud, we stopped at the University of Virginia and James Madison's elegant mansion and estate of Montpelier. We stopped at Appomattox, where General Grant took General Lee's sword in surrender. Then we proceeded on to Charlottesville.

Laura's apartment was situated in the gracious Westminster-Canterbury retirement home—a far cry from Wagon Mound, New Mexico, where I used to call on her. At eighty-three, she had obviously mellowed. Her hair was white and her figure subdued, but she was still quite attractive and lively, and adorned with turquoise jewelry. Her apartment was filled with New Mexico memorabilia. Whether she was disappointed on seeing me at eighty-nine or not, she concentrated her attention on Pamela, which was fine with me.

That evening, we dined at the swank Keswick Hall Club. Fortunately, Laura monopolized the conversation with her lucid recollections of her ranch and New Mexico. She dwelt little on her married life, which initially was in New Mexico, where her husband was involved in oil exploration, and then later led to many stations while her husband was working for the CIA.

The following morning we visited with Laura again. She presented me with her father's favorite bolo tie and gave Pamela a turquoise bracelet. We departed with the satisfying feeling the ice was broken and the romantic fantasy defused.

Later in the summer, Laura satisfied her wish to visit New Mexico. Accompanied by her daughter and son, she spent time with friends in Albuquerque and then came to Santa Fe, where I hosted her for three days. She returned to Virginia reconciled but still longing to move to New Mexico.

After Laura's visit, I returned to my preoccupations with the Cow Creek property, hiking, and writing a Chili Club paper on the BP oil disaster. Fortunately, the spill was contained at the time of my writing, but the topic was timely and warmly received by the "Chileños."

In September Pamela resumed her role as a student. She departed for her course at Christie's in London. Bill accompanied her and helped her get settled before he returned to his technical computer activities in Santa Fe. Perhaps triggered by her absence, or for other unknown reasons, I had a scare on September 28. That morning, after my longtime maid Carol had arrived, I suddenly became dizzy and somewhat nauseated. Carol was alarmed and called 911. The response was quick. An ambulance with two able medics arrived. They checked me and determined I had not had a stroke, but they felt I should be examined further at the hospital. The trip was exciting. The staff at the emergency facilities immediately started the procedures. I spent the day being tested, examined, and observed. The diagnostic conclusion was that I had an inner ear imbalance. I was relieved and impressed by the efficiency and professionalism of the emergency team involved.

Dismissed from the hospital after having been declared fit, I took on a new lease on life. I felt well and energized to finalize my literary journey from 1921 to 2010, and in 2011, my ninetieth year, to commence gratefully recounting this remarkable span of varied and interesting years.

2011: Postscript

Now that my life review has been completed, I have a chance to reflect and render a verdict: I believe I have passed the test, although not with high honors. I did make mistakes but I feel most of my important decisions were generally sound.

I stand ready to continue the race, hopefully successfully. I realize the odds for being selected to rerun the marathon are slim, but I will keep in shape, just in case.

I have learned some important lessons and I'll do my best to put them into practice. I will try not to worry unnecessarily. I will be more decisive. I will not try to control others. I will be optimistic and empathetic. I will be thankful for the amazingly good luck that has been showered on me.

CPSIA information can be obtained at www.ICGtesting.com
Printed in the USA
LVOW12*0311040614

388478LV00001B/1/P